Ana Siqueira do N. M. Teixeira
Paulo Ronaldo S. Teixeira
Roberto A. L. Soares

Use of Foundry Slag in the Production of Structural Ceramics

Ana Siqueira do N. M. Teixeira
Paulo Ronaldo S. Teixeira
Roberto A. L. Soares

Use of Foundry Slag in the Production of Structural Ceramics

ScienciaScripts

Imprint

Cover image: www.ingimage.com

This book is a translation from the original published under ISBN 978-3-330-76386-9.

Publisher:
Sciencia Scripts
is a trademark of
Dodo Books Indian Ocean Ltd. and OmniScriptum S.R.L publishing group

120 High Road, East Finchley, London, N2 9ED, United Kingdom
Str. Armeneasca 28/1, office 1, Chisinau MD-2012, Republic of Moldova, Europe
Managing Directors: Ieva Konstantinova, Victoria Ursu
info@omniscriptum.com

Printed at: see last page
ISBN: 978-620-8-40810-7

Words are important, but gestures are more so;

Words encourage, gestures support;

Words are gone. Gestures remain in the memory forever.

Gestures were and are present In his affection, his attention, his

gaze Understanding, loving, friendly.

Thank you, Paulo, for being in my life!

ACKNOWLEDGEMENTS

Getting this far has only been possible with the help of many. Many people I would like to thank sincerely.

Thank you...

To God, the architect of the universe, for giving me the strength and courage to finish this work, and for showing me the ways out of all the difficult times.

To my parents, José Marreiro and Graça Siqueira, responsible for all my education, wisdom, principles, teachings and values such as dignity, respect and humility.

To my husband Paulo Ronaldo, my unconditional companion. Always by my side, lifting me up and making me believe that I can do more than I can imagine. Because of your companionship, friendship, patience, understanding, support, joy and love, this work could be realised. Thank you for making my dream our dream!

To my little Laura Carollyne, who now inspires me to want to be more than I have been so far!

My sisters for their unconditional love. Drika, who has always been tireless in her expressions of support and affection. Liana, for her companionship and for always being with me to share every moment of joy.

To all my relatives, uncles and cousins. I won't name names, so as not to forget anyone. To the role models I always try to look up to: my maternal grandmother Alba (in memorian) and paternal grandmother Luiza (in memorian), for their eternal unconditional love and for having taught me to be noble in the essence of the word. I miss you all so much!

To my in-laws, especially my father-in-law Joaquim Lúcio Teixeira (in memorian), who, although he is not physically present, I know is celebrating this achievement.

To Professor Dr Roberto Arruda who, with his wisdom and competence, guided this work with great commitment. He was always cordial, helpful and willing to help.

To the Postgraduate Programme in Materials Engineering - PPGEM and to the competent teaching staff for their teachings throughout my Master's degree, especially to Professors Dr Ayrton de Sá Brandim and Dr^{Q} Maria de Fátima Salgado.

To Professor Dr José Milton Elias de Matos (UFPI) who was available to contribute to this work.

Cerâmica Mafrense for supplying raw materials and important information about the ceramics industry in Piauí.

To SENAI-PI for carrying out various research tests through the Clay Technological Testing Laboratory - LETA, in particular to the employee Rui Babosa for the attention and teachings given to me during the course of the research whenever I needed it.

FUNOR - Fundição Nordeste SA, for providing the granulated foundry slag used in this work, especially Mr Jorge, Mr Wilson and Mr Walney for their guidance on the foundry process.

To the Interdisciplinary Laboratory of Advanced Materials - LIMAV - UFPI, for the characterisation analyses of the raw materials used.

And finally, to all those who, directly or indirectly, contributed to the realisation of this dream.

'As long as we're trying, we'll be happy, fighting for the definition of the undefined, for the conquest of the impossible, for the limit of the unlimited, for the illusion of living.

When the impossible becomes a challenge, satisfaction lies in the effort, not just in the final fulfilment." (Gandhi).

SUMMARY

One of the fastest growing economic activities in the state of Piauí is structural ceramics. This is due to the fact that quality raw materials are easily available in the state, as well as the ability of these industries to recycle waste from other industries. This study evaluates the use of granulated foundry slag in ceramic masses for the red ceramics industry, with the aim of not only contributing to the planet's environmental quality by reusing this waste, but also improving the quality of structural ceramics produced in the state of Piauí. To this end, the raw materials basic clay paste and granulated foundry slag were characterised by particle size analysis, mineralogical analysis, chemical analysis, thermal analysis and plasticity analysis. The specimens were formed by extrusion and fired at temperatures of 800°C, 850°C, 900°C and 950°C, with percentages of 0%, 5%, 10%, 15% and 20% concentration in the ceramic mass of granulated foundry slag. Technological tests were then carried out on linear shrinkage, water absorption, apparent porosity, apparent specific mass and mechanical flexural strength. The results showed that the addition of granulated foundry slag to the basic ceramic mass is quite viable, and in some cases there was an improvement in the technological properties, giving higher quality to the ceramic pieces produced, so this could be an alternative for the use of this waste, also contributing to the environmental quality of the planet.

Key words: basic clay mass, granulated slag, reuse, structural ceramics, technological properties.

SUMMARY

CHAPTER 1

INTRODUCTION

With the evolution of industrial processes and the consequent emergence of countless products that quickly became essential, industrial activity has become essential today. Although its importance is indisputable, industrial activity is responsible for generating a very large amount of waste, with different forms and characteristics. As a result, industries have become a major source of solid, liquid and gaseous waste (Ribeiro et al, 2009).

The current panorama of the ceramics industry in Piauí is highly relevant, especially when it comes to the production of structural red ceramics. This is due to the wide availability of suitable raw materials, as well as the lower installation costs that this industry requires to manufacture end products, which are easy to penetrate in the local market.

The natural variability of clay characteristics and the use of relatively simple processing techniques to manufacture red ceramics, such as sealing blocks and roof tiles, make it easier to incorporate waste, improving the quality of the end product, as is the case with the granulated foundry slag investigated in this work (Vieira et al, 2009).

The ceramics industry is one of the most prominent in the recycling of industrial and urban waste, due to its high volume of production, which makes it possible to consume large quantities of waste and which, combined with the physical-chemical characteristics of ceramic raw materials and the particularities of ceramic processing, makes the ceramics industry one of the great options for recycling solid waste. Furthermore, it is one of the few industrial areas that can obtain advantages in its production process by incorporating waste among its raw materials, such as saving high-quality raw materials, which are increasingly scarce and expensive, diversifying the supply of raw materials, and reducing energy consumption and therefore costs (Wender & Baldo, 1998).

The foundry sector is closely related to a country's level of industrial development (Siegel, 1978). In the meantime, Brazil occupies seventh place in the ranking of castings producing countries (American Foundry Society, 2009), indicating its relevance in the global context and consolidating the progress of its industry. Ambiguously, this sector can also be considered a major polluter, as its production processes generate a large amount of waste. These include foundry slag and sand (ABIFA, 2008).

The foundry industry has therefore been looking for alternatives to generating and disposing of its waste. One of them is internal recycling, such as regeneration processes and recovery of the waste used, thereby reducing input consumption. Another is the valorisation of waste as raw material for other processes or activities (Ericksson, 2001; Gorbea, 2001; Naystrom, 2001).

Based on the environmental impacts generated by foundry waste and the trend towards ever-increasing landfill disposal costs, interest in research into better reuse of granulated slag has been growing, making this a material with great economic potential to be exploited in other industrial processes, increasing sustainable development and reducing the polluting potential of the foundry sector in Brazil.

Based on the assumption that the foundry industries are among those that dispose of the most waste, and that the structural ceramics industry in Piauí is one of the fastest growing, there was a need to carry out an experimental study into the reuse of granulated foundry slag in the production of structural ceramics, with the aim of improving their strength and quality, reducing the environmental impact of this waste on the environment.

The work is organised into eight chapters. The first is the introduction, which presents the justification for the work, as well as the importance of recycling industrial waste; the second chapter sets out the objectives to be achieved with the research; the third chapter contains information from the literature on the structural ceramics industry in Brazil and Piauí, as well as information on the raw materials used in the sector, the manufacturing process and the main structural ceramics products; the fourth chapter is made up of information from the literature on the foundry industry, production processes and foundry waste; the fifth presents an approach to industrial waste, from classification, recycling and the recycling of solid waste with incorporation into structural ceramic mass; the sixth chapter presents the experimental procedure adopted; the seventh presents the results achieved and the respective discussions; finally, the eighth presents the main conclusions based on the results of the research.

CHAPTER 2

OBJECTIVES

> *General Objective*

- The aim of this work is to present the technical and environmental feasibility of incorporating granulated foundry slag into the production of structural ceramics.

The Specific Objectives

c Characterise the samples of granulated foundry slag and basic mass

of the clays used in the research through the analyses of Granulometry (GA), Plasticity Analysis (PI), X-Ray Diffraction (XRD), X-Ray Fluorescence (XRF), Gravimetric Thermal Analysis (TG);

- Analyse and interpret information on the technological properties of: Loss on Firing (LF), Linear Shrinkage (LR) on Drying and Firing, Water Absorption (WA), Apparent Porosity (AP), Apparent Specific Mass (ASM) and Breaking and Flexural Tension (BFR) on Drying and Firing of standard ceramic mass formulations and those with the addition of different levels of slag at different temperatures;

- Evaluate the possible improvements in the quality of structural ceramic pieces formulated using this waste;

- Developing environmentally safe and effective structural ceramic materials for the construction industry;

- Improving the planet's environmental quality by reusing waste in new industrial processes.

CHAPTER 3

THE STRUCTURAL CERAMICS INDUSTRY

Historically, the ceramics sector was known as the activity of producing artefacts from moistened clays which, after drying, were moulded and subjected to high temperatures, giving them rigidity and resistance through the fusion of certain components of the mass.

Currently, the ceramics industry has a wide variety of products and production processes, with different types of establishments coexisting, with different characteristics in terms of production levels, product quality, productivity rates and degree of mechanisation.

In Brazil, the ceramics industry is mainly concentrated in the south-east and south, where the country's largest ceramics centres are located. However, other regions have shown some development in this industry, especially the northeast, mainly due to the existence of raw materials, viable energy and a developing consumer market. Production in the north-east of the country accounts for 21 per cent of national production (BNB, 2010).

The ceramics sector can be divided into the following segments: red or structural ceramics, cladding materials, refractory materials, sanitary ware, tableware and porcelain, porcelain electrical insulators, artistic ceramics (decorative and utilitarian), ceramic water filters for domestic use, technical ceramics, thermal insulators, lime and cement, the latter due to their specific characteristics, which are little accounted for in the sector (ABC, 2002).

Red or structural ceramics are a type of ceramic characterised by the red colour of its products, which are used in civil construction, such as bricks, blocks, tiles, pipes, ceiling slabs, hollow elements, light expanded clay aggregates and others. These are products that offer good durability, thermal and acoustic comfort and low cost, among all those involved in the construction production chain (ABC, 2002).

3.1 - Brazilian Industrial Profile

Most of the companies that make up this segment are micro, family-run or small and medium-sized businesses that generally use technologically outdated production processes. They are spread throughout the country, with large concentrations of this type of industry due mainly to the availability of raw materials (SEBRAE, 2005).

The ceramics factories are preferably located in regions close to deposits of clay, the main raw material used to make their products. This gives red ceramics a special **"inland"** industry configuration, **since** most of the enterprises are located in rural areas in search of clay deposits and are responsible for a significant portion of the local labour force (Bezerra, **et al.**, 2005).

Characteristics of this type of industry include family management, size linked to the exploitation of deposits, low start-up costs, low barriers to entry for new competitors and technological processes that are behind the standard in other countries.

Due to the production of low value-added items and transport by road, sales in this segment are destined for markets not far from the industrial unit (SEBRAE, 2005).

Surveys show that the red ceramics industry stands out on the national scene, accounting for 7.3 per cent of the Gross Domestic Product (GDP). It is made up of approximately 7,430 companies in Brazil, directly employing 293,000 workers and generating 1.25 million indirect jobs, as well as annual sales of R$6 billion, according to ANICER - the National Ceramic Industry Association (2010). With regard to this type of industry, it can also be seen that its growth is linked to that of the construction industry. According to the Mining and Energy Yearbook (2009), the non-metallic sector will see a greater demand for these products, due to world events taking place in Brazil over the next few years.

The average productivity of the Brazilian pottery-ceramics segment is around 15,000 pieces/worker/month, which varies greatly depending on the region. The low productivity compared to other countries shows the need to modernise the segment in Brazil (Ministério de Minas e Energia, 2009). It is also estimated that the total consumption of tiles in the country in 2008 was around 75 billion tiles, of which 16.5 billion were consumed in the Northeast region.

The most important factors in the Brazilian structural ceramics production process are labour, raw materials and energy consumption. Labour costs play a significant role in the cost of production, which is generally unskilled and has a high turnover rate. Energy costs play a significant role in the costs of the red ceramics industry. The biggest cost is the purchase of firewood and sawdust used in the drying and firing process, while electricity is used for other equipment (SEBRAE, 2005).

The need for investment in the ceramics sector to improve quality and productivity is a growing concern. The materialisation of this trend is still slowly taking place, through new management techniques and, above all, the introduction of more up-to-date and efficient processing, as seen in some structural ceramic block and roof tile factories (Motta et al., 2001).

3.2 - Piauí's ceramics sector

The ceramics industry in Piauí is of significant importance to the economy and is a major generator of jobs in the state. It is a regional benchmark for quality ceramic products, especially roof tiles, which are among the best in the country. The outlets for the state's production reach the states of Pernambuco (mainly Petrolina), Pará, Ceará (Serra da Ibiapaba) and Bahia (mainly Juazeiro) (INT, 2012).

The red ceramics sector in the state has 60 companies (20 potteries) in full operation, 10 of which are unionised, together producing around 43,000 thousand/month of blocks (83%) and tiles (17%) and providing 2,500 direct jobs (42 employees/company). Demand for clay is around 96,000 tonnes/month and actual production 82,000 tonnes/month, giving a ratio of around 1,900 kg/millimetre. The average production per company is 716 thousand/company.month, which is above the average for the Northeast, with 67% of them between 200 and 1,000 thousand/month and 18% above 1,000 thousand/month. The largest production centres are, in order, the municipalities of Greater Teresina (42% of state production), Campo Maior (10%) and Parnaíba (4%). In Greater Teresina, 32 ceramics factories operate, 24 of them in Teresina, totalling a production of 18,000 millimetres/month (average of 563 millimetres/company.month) (Schwob, 2007).

Piauí's most important ceramics centre is located in the region around the capital, Teresina, and is one of the most important in the Northeast. This region has large deposits of quality clay for red ceramics, mainly on the banks of the Parnaíba and Poty rivers, and is responsible for 70% of the state's clay extraction. Ceramics located in the neighbouring city of Timon, in the state of Maranhão, also benefit from these deposits (CEPRO, 2005).

In around half of the companies, the clay deposit is located less than 5 kilometres from the factory and 40% between 5 and 10 kilometres. Only 25% of the companies store clay for more than a year and 55% for a maximum of 6 months. Around 60% carry out artificial and natural drying, with only 14% using semi-continuous dryers fitted with thermo-hygrometers and thermostats (INT, 2012).

The ceramics sector in Piauí, spread across the state in mostly small and medium-sized companies, manufactures products such as tiles, bricks, blocks, flagstones, ceramic flooring, ornaments, utilities and others. The state's production is uncertain, as many of these companies are not legalised (Soares, 2008).

In general, the ceramics industry in Piauí is made up of small companies that offer differentiated products with good prices and market penetration, and a large number of small producers who concentrate their activities on rudimentary exploration and the manufacture of products of low quality and price. This reality tends to change given the potential of the state's mineral resources and the prospects for development (SINDICER-PI, 2008).

The most widely used type of kiln is the intermittent reversible circular vault kiln, which accounts for around 80 per cent of the production park. Hoffman-type kilns account for 11 per cent and the remainder is made up of kilns of various types, such as Federico, Paulistinha and Campanha (INT, 2012).

The state's ceramics factories predominantly use firewood and vegetable waste, mainly babassu bark, as fuel for their kilns, with a demand of around 50,000 m^3/month (firewood and waste). Firewood, which is more expensive at around R$25.00/m3, accounts for around 60% of demand

(27,000 m3/month) (INT, 2012).

One of the biggest difficulties facing the state's ceramics industry is preserving the environment. The mining of the raw material deforests and digs up the soil, forming huge holes that later have to be used to breed fish in captivity or are left with no practical use. The use of firewood for firing ceramics is another factor that causes environmental damage, with deforestation in the area of origin and pollution caused by the polluting gases emitted into the atmosphere during firing (CEPRO, 2005). In addition to the environmental impacts, the ceramics sector in Piauí faces other difficulties such as a low level of technology, generating significant losses in production and quality; an unskilled workforce with a low level of education (Soares, 2008).

3.3 - Ceramic raw materials

The fundamental raw material for the red ceramics manufacturing process is clay. Clays are predominantly made up of clay minerals (phyllosilicates), and their most common types are made up of tetrahedral (T) sheets of silicon and octahedral (O) sheets of aluminium, and less often magnesium and/or iron. They are structured in a 1:1 (TO) or 2:1 (**T2O**) ratio. In addition to the structural arrangement, the basal spacing of these units characterises the clay minerals in the various groupings, with the kaolinite, illite and smectite groups standing out as the most important for ceramic use. Other minerals occur with the clay mineral particles, generally in the silt **(0.002 mm< <t> > 0.62 mm) and sand (d>** > 0.62 mm) fractions. In these larger particle sizes, the most common and abundant mineral is quartz, followed by micas, feldspars and opaque minerals (CTEM, 2008).

The clay minerals most easily found in clays used in the red ceramics industry are kaolinite, illite and montmorillonite (Monteiro, 2009). Montmorillonite is characterised by extremely fine particles and its general formula is $(OH)_2(Al, Mg, Fe)_2(Si_2O_5)_2$. Its main characteristic is its ability to absorb water molecules between the layers. As a result, clays rich in this clay mineral have a strong tendency to cause drying cracks, as well as being highly plastic (Grun, 2007).

Essentially montmorillonitic clays are undesirable in the manufacture of products such as ceramic blocks and tiles, due to their difficult "workability", i.e. they reduce the productivity of the machinery due to their high plasticity and make it difficult to dry the pieces due to the degree of packing of the particles due to their small grain size. In the case of these excessively plastic clays, i.e. predominantly montmorillonitic, it is essential to use less plastic clays in order to provide the ceramic mass with satisfactory moulding, drying and firing conditions (Monteiro, 2009).

Kaolinite is the most common clay mineral found in clays. Its basic structure is $Al_2O_3.2SiO_2.2H_2O$. Clays made up essentially of the clay mineral kaolinite are the most refractory. Illites have part of the silicon replaced by aluminium, as well as containing more water between the layers and having part of the potassium replaced by calcium and magnesium. Their exact chemical composition is difficult to determine because they are always contaminated with impurities that are difficult to eliminate. Because they have potassium in their structure, they have good resistance after sintering (Grun, 2007).

Thanks to clay minerals, clays develop a series of properties in the presence of water, such as plasticity, porosity, wet mechanical resistance, linear drying shrinkage, compaction, thixotropy and viscosity of aqueous suspensions, which explain their wide variety of technological applications (ABC, 2002). Excessively plastic clays are rich in alumina (the plastic part) and lean clays are rich in silica (the non-plastic part). Greater plasticity is also strongly influenced by the fine granulometry and organic matter present in the clay (Rêgo, 2010).

According to Motta et al. (2004), clays are classified into two main groups based on their firing colour and mineralogy. These two subgroups have links to industrial application sectors, where: a) reddish firing clays are used as raw materials for red and structural ceramics, ceramic tiles (includes mainly dry kiln products and secondarily, reddish extruded floor tiles and wet kiln products), light aggregate, vases, pots, utilitarian and decorative pieces with a reddish body; b) light-fired kaolinitic clays (plastic clays and kaolin for white ceramics) are used as raw materials in sanitary ceramics, wet ceramic plates, technical porcelain, utilitarian and artistic porcelain and earthenware. Some of these clays can also be used in the manufacture of silica-alumina refractories and possibly mullite, but refractory clays will not be dealt with in detail.

The main characteristic of common clays for red or structural ceramics is their reddish firing colour. This property is due to their high content of total iron oxide, which generally exceeds 4% (Facincani, 1992).

For the manufacture of bricks, clays must be easily moulded, have medium to high tensile strength values, a red colour after firing, few cracks and warping. It must also have high iron contents and low alkaline and alkaline-earth element contents. For the manufacture of roof tiles, clays must have adequate plasticity for moulding, high flexural tensile strength when dry, so as to allow handling during the manufacturing process. After firing, it must have low apparent porosity and low water absorption and be free of cracks and warping after drying and firing (CETEM, 2008).

The study of clay raw materials used in the structural red ceramics industry is aimed at finding information that can help in the development of products and processes. The result can be reflected in better quality bricks and tiles, either through changes in the formulation of the mixtures or improvements in the manufacturing process through control of the properties of the raw materials (Grun et al, 2005). Knowing the raw material you are working with means knowing its drying and firing shrinkage, its plasticity, its granulometry, the amount of water it requires for extrusion, its tendency to dry cracks, as well as its resistance after firing.

In the manufacture of ceramic tiles, it is common to mix two or more materials for the composition of the mass (Vieira **et al**, 2001), as well as additives and water or another medium (ABC,

2011b). It is difficult for a single raw material alone to provide the best possible internal structure and quality for the ceramic mass (Pérez **et al.**, 2010). This clay formulation generally seeks, empirically, an ideal composition of plasticity and fusibility, so as to enable easy compaction and mechanical resistance during firing (Motta **et al**, 2001). Traditional ceramic masses are named according to particular characteristics such as colour, texture and shape (Correia, 2004).

In the formulation of ceramic **clay**, at least two **types of clay** are usually mixed**: a "fat" clay, which is characterised by high** plasticity, fine granulometry and a composition essentially of cl**ay** minerals; and a **"lean" clay, which is rich in quartz and less** plastic, and can be characterised as a plasticity-reducing material (Motta, **et al.**, 2001).

3.4 - Ceramic Processing

The ceramics industry is characterised by two distinct stages: the primary stage (which involves exploiting the raw material - in this case, clay) and the transformation stage (to produce the final product) (SEBRAE, 2008). The manufacturing process basically goes through six stages: extracting the raw material, seasoning, preparing the dough, shaping the pieces, drying and firing, according to the flowchart in Figure 1, illustrated below.

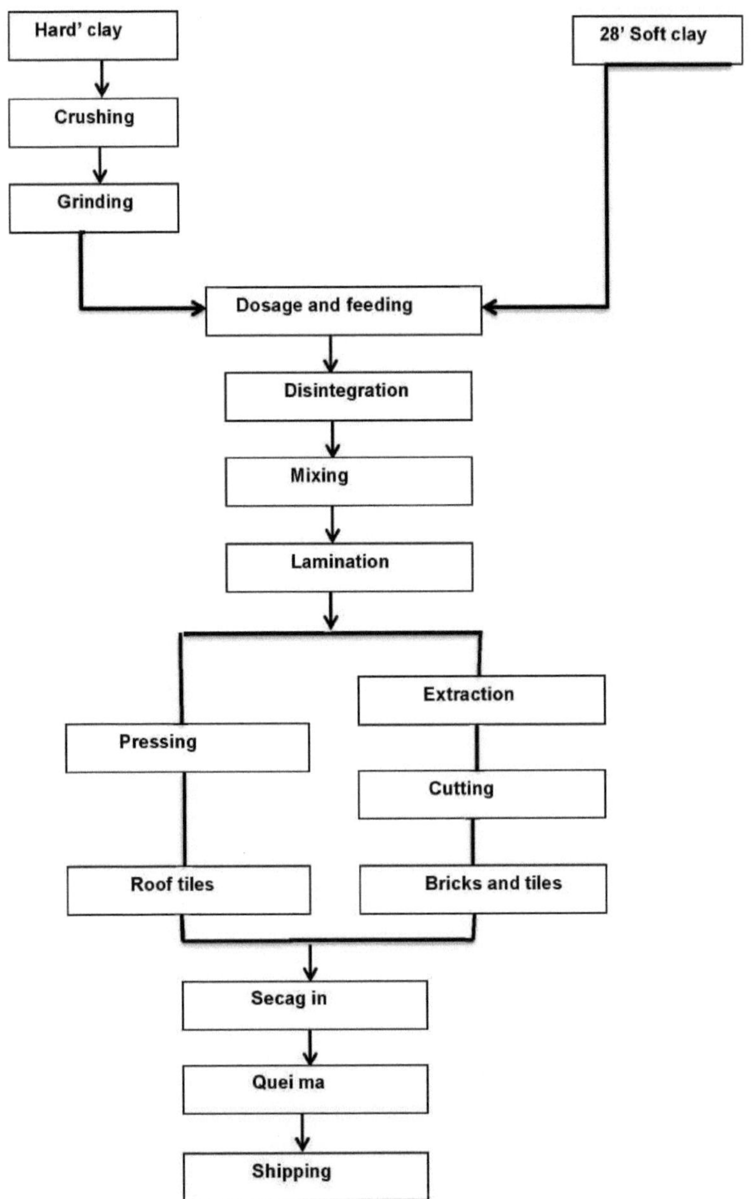

Figure 1 - Red ceramics production chain
Source: Reproduced from ABC (2002). Technical information: manufacturing processes.

The quality of the structural ceramic product is linked to the stages of the manufacturing process, from the selection of raw materials to the storage of the final product. During these various stages, problems can occur that cause the final product to be distorted, reducing its quality. Good management of all the phases helps to avoid waste and losses and to guarantee quality and compliance with the technical standards in force. And a good quality end product not only attracts consumers, but favours the development of the sector as a whole, as it increases competitiveness between companies in the sector (Silva, 2009).

3.4.1 - Raw Material Extraction

Most of the raw materials used in the production of structural ceramics are natural and are found in deposits scattered throughout the earth's crust. In general, clays are extracted in the open as shown in Figure 2, preceded by preliminary studies (chemical analysis tests, X-ray diffraction, plasticity, among others) for the technological characterisation of the material. The auxiliary means available for extracting and transporting the clays to the storage site range from rudimentary equipment to large mechanised equipment (Morais, 2007).

Figure 2 - Raw Material Extraction

3.4.1 - Stocking: Seasonal

Seasoning consists of storing the raw material in the company, as illustrated in Figure 3, for a period that can vary from six months to two years, with the main aim of improving the plasticity of the clays, washing out soluble salts, decomposing organic matter and reducing the tensions caused by the breaking of chemical bonds, facilitating extrusion moulding (SENAI, 2006). In addition, the seasoning period prevents the expansion of the pieces immediately after moulding, with the occurrence of deformations, cracks and ruptures in the products during the drying phase, and the development of gases during firing (Petrucci, 1998).

Figure 3 - Seasoning (Stockpiling)

3.4.3 - Dough preparation

After seasoning, the raw material is transported to the feeder bin, where, if necessary, dosing is carried out, which must strictly follow the previously established proportions. The dosed mixture is taken to the disintegrator, where the clods are broken up. The material is then transported to the mixer, where it is homogenised and finally adjusted by adding water where necessary. Finally, the mixture is transferred to the rolling mill, which has the task of pre-laminating and completing homogenisation (Morais et al, 2006). Figure 4 shows the stages of dough preparation, from the feeder box to the rolling mill.

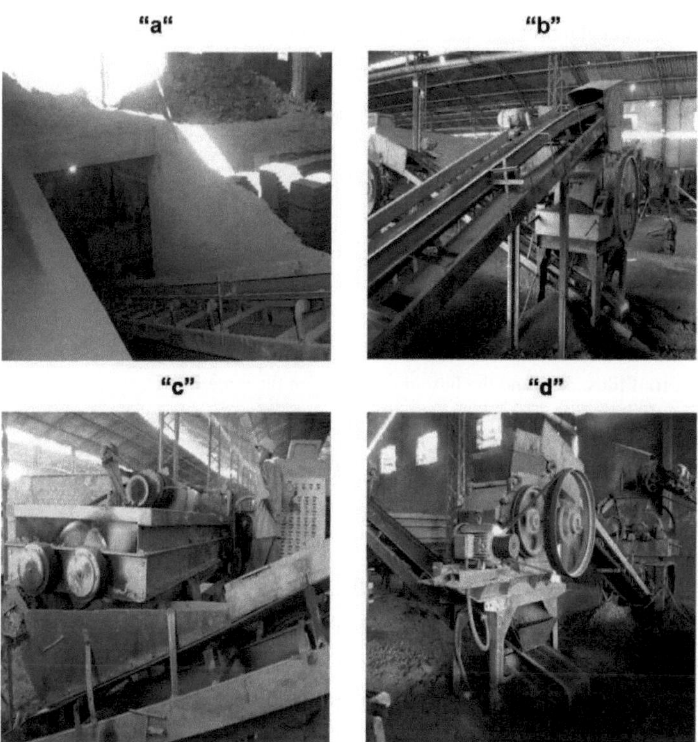

Figure 4 - Ceramic mass preparation stages: "a" - Feeding bin, "b" - Disintegrator, "c" - Mixer and "d" - Rolling mill.

The formulation of structural ceramic clay is generally done empirically, with an ideal composition of plasticity and fusibility to provide good workability and mechanical firing resistance. The clay is usually prepared by mixing a "fat" clay, which is characterised by high plasticity and fine granulometry, with a "lean" clay, which is less plastic and coarse granulometry. The mixture is then moistened to an average content of 20% and homogenised to form the ceramic products (Motta, et al., 2001).

Large grained clays (up to 600mm) must be pre-crushed until they are compatible with the dimensions of the primary mill's feed nozzles (up to 200mm). Depending on their hardness, workability and grindability, clays can be categorised on the Mohs scale as Hard (6-7 Mohs), Semi-hard (5-6 Mohs) or Soft (less than 4 Mohs). The ideal condition for grinding is that the clay is categorised as hard or semi-hard and has a moisture content of no more than 18%. Otherwise, use can be made of the various preparation methods available, whether mechanised or not (Petrucci, 1998).

It is important to control the milling process, as too little milling can cause coarse particles in the mass and lead to dimensional destabilisation, making the mass less reactive and compromising the product's technological properties. Very fine granulometry can lead to the following problems: weak cohesion of the dough being formed; low mechanical strength of the pressed and dried product; longer drying and firing cycles and the risk of defects in the final product (Manfredini & Schianchi, 2009).

3.4.4 - Parts forming

The method of forming or moulding must be chosen according to the product required (sealing blocks, structural blocks, roof tiles, etc.) and the humidity present in the material.

The methods used to form blocks and tiles are extrusion and pressing. Extrusion is the most widely used method in the structural ceramics industry, as it is more economical, but generates lower value products. Pressing, a more expensive method, is used for some types of tiles with higher added value.

According to Morais et al (2006), extrusion is the most widely used forming method in the ceramics industry and consists of moulding the ceramic mass into a plastic but rigid paste in equipment called an extruder (or maromba) (Figure 5), made up of an auger that compresses the mixture as it passes through a perforated steel plate. The material is then chopped up, falling into a vacuum chamber where deaeration is carried out.

Just below this chamber, another auger forces the dough through the nozzle, which prints the predefined shape, forming a continuous column, which is cut to the desired length.

Figure 5 - Maromba used for extrusion in the ceramics industry with ceramic mass coming out of the nozzle.

According to Norton (1973), in the dry pressing method the moisture content used is very low (less than 10%) and the pressure exerted is quite high. This method is generally used for moulding bricks, tiles and refractories, as well as better quality bricks and tiles. The method consists of placing the granulated mass with a lower moisture content in a mould made of rubber or other polymeric material, closing it hermetically, and introducing it into a chamber containing a fluid, which is compressed and consequently exerts a strong pressure evenly on the mould (SENAI, 2006).

3.4.5 - Drying

After forming, the drying stage begins. The drying process is a very important operation in the manufacture of structural ceramics, requiring special care to ensure that the water contained in the products is slowly and evenly eliminated throughout the ceramic mass to avoid possible defects in the piece such as cracks, warping or breakage (Norton, 1973).

At this stage, it can be seen that as the water evaporates, the raw piece densifies and, consequently, mechanical strength increases. This is attributed to the packing and attraction between the particles, which increases the bonding forces between them (Oliveira et al, 2000). According to Van Vlack (1973), the water deposited between the particles of ceramic tiles must be eliminated during the drying stage, as this water causes undesirable cracks in the products at the start of the firing stage. The higher the moisture content in a piece, the greater its shrinkage.

Beltran et al (1995) also points out that the water of conformation can be divided into two types: the first is called interstitial water. The second is called free water or plasticity water. Interstitial water is related to the water needed to fill the pores of the particles. Plasticity water, on the other hand, is located between the clay particles, separating them and facilitating workability in the forming process. It is this last type of water that is responsible for the shrinkage of the pieces.

Drying can be natural (exposing the pieces to the open air) or forced (intermittent or continuous dryers) (Vieira et al., 2003). When drying is natural, its duration depends on the conditions of the atmospheric air (temperature and relative humidity) and the ventilation in the room, and can last up to six weeks. Artificial drying is carried out in kilns at a temperature of between 80 °C and 110 °C. The drying time depends on the characteristics of the raw material, the shape of the pieces and the type of dryer, but there is an average variation of between 12 and 40 hours (Pauletti, 2001).Figure 6 shows the natural drying of brick blocks carried out in the ceramics industry.

Figure 6 - Natural drying used in the ceramics industry

3.4.6 - Burning

During firing, ceramic products acquire their final properties, such as mechanical strength and colour, among others. After drying, the pieces are subjected to heat treatment at high temperatures, which for most products is between 800 °C and 1700 °C (ABC, 2001b).

There are four phases in this process: a) preheating is characterised by gradual heating to remove residual water, without causing defects in the ceramic piece caused by differential contractions during the expulsion of the remaining moisture, over a period of 8 to 13 hours, reaching up to 650 °C; b) heavy firing or boiling, starts at around 650 °C and can be increased at a faster rate up to 950 °C or 1000 °C. This is when the chemical reactions take place that give the ceramic body its characteristics of hardness, stability, resistance to various physical and chemical agents, as well as the desired colour; c) plateau, is the maximum firing temperature maintained for a certain period of time, in this way the chamber maximises the temperature gradient throughout the kiln; d) cooling, is carried out gradually and carefully to avoid cracking, through the chimney or by using the heat for the dryers, over a period of around 38 to 50 hours (SENAI, 2006).

According to Bauer (2000), important chemical reactions take place during this phase, giving the blocks final characteristics such as mechanical strength and colour, among others. The main aspects that influence firing are: temperature, firing time, heating and cooling speed, ambient atmosphere, type of kiln and fuel used, among others.

The kilns used to fire blocks can be classified as periodic (or intermittent) and continuous. Intermittent inverted flame kilns are the most commonly used in the Brazilian ceramics industry. In these kilns, the hot gases enter the upper part of the firing chamber, pass through the pieces and exit through grates that lead to underground channels connected to a chimney, which promotes the exit of the gases (Oliveira, 2002).

Continuous kilns are thermal equipment in which the firing cycle is carried out without interruption, i.e. while the firing of the ceramic blocks in one carriage comes to an end, another is being started, without discontinuity (Morais **et al**, 2006).

Figures **7 "a" and "b" show** the types of intermittent and continuous kilns used in the ceramics industry.

Figure 7 - Types of kilns used in the ceramics industry "a" shows the intermittent vault-type kiln and figure "b" a continuous tunnel-type kiln.

3.5 - Main structural ceramics products.

The main products manufactured by this sector, which are primarily destined for the construction industry, are: solid and perforated bricks (in various sizes), roof tiles of various models, ceramic sealing and structural blocks, ceiling and floor slabs, hollow elements and floor tiles (Varela, 2004).

3.5.1 - Structural ceramic bricks/blocks

Ceramic sealing blocks are components of structural masonry. They have prismatic holes perpendicular to the faces that contain them, but in this case, structural ceramic blocks are produced so that they can be laid with vertical holes. They make up external or internal masonry

which do not have the function of resisting vertical loads other than the weight of the masonry of which it is part, as described in NBR 15270-2 (ABNT, 2005).

Blocks or bricks are the very essence of masonry, as they are responsible for the main properties in masonry. Their main characteristics are: appearance - the visual characteristics that are of structural or aesthetic interest; dimensions - it is through the dimensions that the uniformity of the blocks or bricks is sought, which is important for the modulation of the structural masonry; water absorption - a property that is directly related to water permeability and adhesion between materials; compressive strength - a mechanical property that defines the compressive characteristics of each unit and is important in defining the variables to be considered in the structural design (ABNT, 2005). Figure 8 shows examples of ceramic blocks with vertical and horizontal holes.

Figure 8 - Examples of ceramic blocks with vertical and horizontal holes

3.5.2 - Roof tiles

NBR 15310 classifies ceramic tiles as components intended for the assembly of watertight roofs for discontinuous application (ABNT, 2005).

These materials, like the blocks, are made up of clays rich in illite and montmorillonite; the raw material, however, is better selected and the ceramic mass better prepared in order to achieve characteristics compatible with the geometry and use of the product, i.e. good dry mass strength, tiles with high flexural strength, low porosity and excellent conformity and aesthetics. The main requirements that must be met include: no cracks, exfoliations, breaks or burrs that hinder perfect coupling between the tiles; adequate and uniform firing; low weight; low water absorption and impermeability; regularity of shape, dimensions and colour within the established standard; surface without roughness; fine edges; low porosity and resistance to bending (ABNT, 2005). Figure 9 shows the types of tiles manufactured in the ceramics industry.

Figure 9 - Examples of roof tiles manufactured in the ceramics industry.

CHAPTER 4

FOUNDRY INDUSTRY

The casting process consists of manufacturing metal parts by filling a mould with liquid metal, the cavity of which has dimensions similar to those of the part to be produced. Monticelli (1994), in a more technical definition, states that casting consists of "the preparation, melting and refining of metallic inputs, their pouring into moulds (by gravity, pressure, centrifugation or vacuum) and the cleaning and finishing of the raw parts thus obtained".

According to Rossitti (1993), although there is no consensus, it is believed that the casting process has been known since 5000 BC, when cast copper objects were made using chipped stone moulds. Although iron ore is found in abundance in nature, the first known iron casting is considered to be relatively recent, dating back to 600 B.C. It is a 275kg tripod produced in China (Loper et al, 2003). Other sources indicate that the iron casting process was known before this, around 1000 BC, where the Chinese were already producing cast iron pieces at higher temperatures, obtained in coal furnaces blown by bellows (Ribeiro,2008).

The first iron castings had very low resistance to fracture. Only later was charcoal introduced during the melting process, giving the final piece greater strength. Innovations in the production method were made in the 17th century by means of incarbonisation, which consists of adding carbon to iron - which would later give rise to steel. The steel casting process dates back to 1740 and is attributed to the Englishman Benjamin Huntsman (Ribeiro, 2008).

In Brazil, the first foundry house appeared around 1580, in São Paulo, and was intended for smelting gold extracted from the mines of Jaraguá and the surrounding area. During the 18th century, many foundries were set up in Minas Gerais, Goiás, Mato Grosso and Bahia. Iron was smelted from the 17th century onwards and, in the last days of its rule, the Portuguese crown even built some blast furnaces in the colony. The demand for railways and ports fostered the performance of foundries for a long time, so that the railway companies' repair yards and shipyards came to have the best-equipped metallurgical workshops in the country (Bethell, 2002).

Because the cast product is basic to most production chains, the presence of the foundry activity is a factor of great relevance to the industrial development of countries (Siegel, 1978) and can therefore be used as an indicative parameter of this development.

4.1 - The Foundry Industry Market in Brazil.

The foundry industry produces iron, steel and non-ferrous alloy castings. There are approximately 1,400 foundry companies (mostly nationally-owned and small and medium-sized), totalling around 60,000 employees and a turnover of 10.7 billion dollars in 2009 (Carmelio et al., 2011).

Production in the sector has grown rapidly in recent years, exceeding 3 million tonnes since 2006 (Carmelio et *al.,* 2011). According to data from ABIFA (2011), Brazilian production of castings in 2009 was around 2,297 thousand tonnes.

This sector's high turnover is due to its intense use of labour and self-sufficiency in raw materials, thus contributing to its independence from the foreign market (Fagundes, 2010). This scenario provides a significant number of direct and indirect jobs in its production chain, which with growing exports and basically no imports, contributes to our country's balance of trade always achieving a positive result.

On the world stage, Brazil is the 7th largest producer of castings (based on 2008), surpassing countries such as Italy, France and Korea, behind China, the United States, Russia, India, Germany and Japan (Carmelio et *al.,* 2011).

Domestically, the foundry industry is present in all regions of the country, with a greater concentration in the south-east and south (Assunção et *al.,* 2007). This means that the foundry, as an indicative parameter for business investment, is present where the major industrial centres are located. In the meantime, the southern region ranked 2nd in national castings production (ABIFA, 2009), showing its representativeness on the national scene. The state with the highest production in 2008 was São Paulo, which reached around 1.16 million tonnes, corresponding to 34.6% of national production (Fagundes, 2010).

The main consumers of domestic castings are the Brazilian automotive industry (58 per cent), exports (13 per cent), capital goods (13 per cent), infrastructure (6 per cent) and the steel industry (2 per cent) (ABIFA 2011).

The consumption of castings by the automotive sector in Brazil is due to the presence of automakers from Asia, the USA and Europe in the country, combined with the large number of vehicles in its territory, around 25.6 million (the tenth largest vehicle fleet in the world), driving the market for manufacturers of automotive components and parts (Carmelio et a/., 2009).

4.2 - Cubilot Furnace Casting Process

The casting process consists of melting a metal which, in a liquid state, is poured in the quantity required to fill a mould and which, when it solidifies, generates a piece with the desired shape (Campos Filho, 1978).

Granulated foundry slag, the waste generated from the production of cast iron and the subject of this work, is obtained through a melting process in a cubilot furnace.

The cubilo furnace is a melting furnace which, using ferrous metal raw materials, makes it possible to obtain (by heating and physical-chemical reactions) cast iron of a specific composition, production and temperature. In general, the cubilot is a vertical furnace made of sheet steel lined (internally) with refractory bricks; the top is open and the bottom consists of a pair of cast iron doors whose purpose is to allow the removal of unconsumed coke and unmelted metal after each processing (Vallina, 1998). Figure 10 illustrates the cubilot furnace used in the iron smelting process.

Figure 10- Cubilô smelting furnace.

According to Caspers (1999), the result of melting in the cubilo furnace is determined by the charge and combustion processes and also, significantly, by the quantity and composition of the slag. The same author also considers that in cubilot furnaces, the retention and gathering of slag with oscillating basicity causes considerable dispersions in the carbon, silicon, manganese and sulphur contents. The final liquid iron content is the result of several reactions, which are partially opposite.

Figure 11 shows the stages of the cubicle furnace casting process.

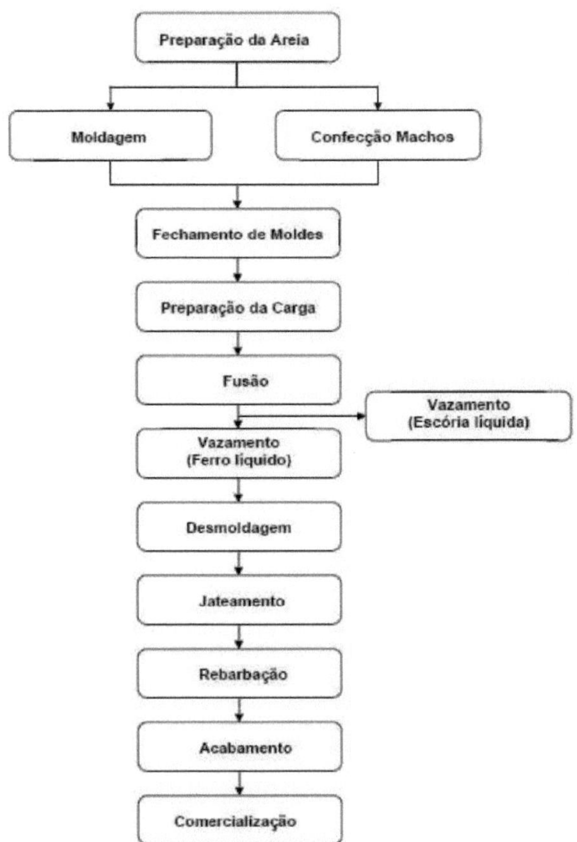

Figure 11 - Flowchart of the foundry process with waste outputs. Source: Adapted (Oliveira, 1998).

4.2.1 - Model making

Moulds, devices that receive the molten material in order to obtain the part, are obtained by moulding, usually in green sand, in a process in which the external shape of the product to be obtained is transferred to the sands by compacting them on a model, usually bipartite, each in a casting box (Campos Filho, 1978; Carnin, 2008). Mould making basically consists of a cavity left in a material by the model of the part to be cast.

The cores, devices with the function of forming voids, holes and recesses in the part, are generally made of chemically bonded sand and are responsible for giving the internal shape of the cast product to be obtained (Kondic, 1973). They are placed in the moulds before they are closed to receive the liquid metal.

4.2.2 - Metal Melting and Leaking

In general terms, once the mould has been cast, the two halves of the mould are joined (with or without the inclusion of cores, depending on product requirements) and the liquid metal is poured into the mould, filling its entire cavity (Campos FIlho, 1978; Kondic, 1973). It is at this stage of the process that the granulated casting slag is removed.

4.2.3 - Demoulding

It is the removal of the part from the mould after the metal has solidified, and can be done manually or by mechanical processes. After the metal has solidified, all the material contained in the mould boxes is subjected to vibration in order to separate the castings from the casting sands (Campos Filho, 1978; Chegatti, 2004).

4.2.4 - Deburring and cleaning

This involves the removal of feed channels, masses and burrs that form during casting, as well as the removal of mould deposits on the casting, usually by means of abrasive jets (Kondic, 1973).

4.3 - Granulated Foundry Slag

Slag can be defined as a non-metallic product consisting essentially of oxides and other bases that develop in the furnace during the iron melting process. Physical characteristics such as density, porosity and particle size depend on the type of slag cooling and its chemical composition (Kalyoncu, 2000).

According to Moraes (2000), the three main components of cubic furnace slag are: alumina (Al_2O_3), which comes from the wear of the refractory; silica (SiO_2), which also comes from the refractory, ash and sand adhering to the filler materials; and finally, calcium oxide (CaO), which comes from limestone. The CaO/SiO2 ratio is what gives the slag generated its acidity.

Based on the above, we can see the strong influence of coke ash and furnace refractory on slag composition. Therefore, acidic foundry slag normally contains 40 to 30 per cent SiO_2, 10 to 20 per cent Al_2O_3, 25 to 38 per cent CaO, 1 to 8 per cent FeO and 1 to 5 per cent MnO (Mahan, 1984).

During the cubic furnace casting process mentioned above, the slag is poured into the liquid through its own channel, located on the side of the furnace, at a temperature of approximately 1500 °C, as shown in Figure 12.

Figure 12- Leakage of liquid slag from the Cubilô furnace.
Source: (da Silva, 2006)

In this channel, through a continuous flow of water, the casting slag undergoes sudden cooling, which results in its granulation, disintegrating into small particles similar to coarse sand. Most foundries do not have tanks for depositing slag, so it is deposited in open pits until it solidifies. Figure 13 illustrates granulated foundry slag after the cooling process.

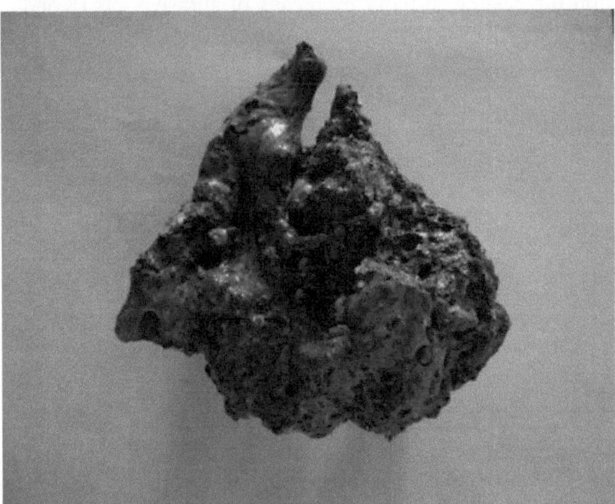

Figure 13 - Granulated Foundry Slag

CHAPTER 5

INDUSTRIAL WASTE

Industrial activities transform raw materials extracted from the environment into products intended for use by the end consumer or into products that can be applied to other production systems. Normally, in addition to the end product, secondary products called industrial waste are produced in industrial activities, as illustrated in Figure 14.

Figure 14 - Industrial Waste Production Flowchart

According to CONAMA Resolution 313/2002, industrial waste is all waste resulting from industrial activities and found in solid, semi-solid, gaseous and liquid states, the particularities of which make it unfeasible to discharge them into the public sewage system or bodies of water, or require technical or economically unfeasible solutions in view of the best available technology (Brazil, 2002).

In the beginning, as industrial activities were fairly rudimentary, the amount of waste disposed of was fairly insignificant and caused little impact on the environment. With the sudden change in industrialisation and population growth, the amount of waste generated and the types of waste, both inert and hazardous, biodegradable and non-biodegradable, recalcitrant and xenobiotic, have increased (Bidone, 1999), which has had an impact on the environment.

The treatment given to industrial waste is usually not the most environmentally friendly. It is only ever disposed of in a way that ranges from open dumping to landfill sites designed for its proper conditioning. While inputs and products are classified based on quality specifications linked to their respective applications, waste is traditionally grouped according to the receiving bodies: water, air and soil, depending on the disposal with which it is associated. Given these associations in production processes, it is important to implement improvements in the search for zero waste and not just think about how it will be disposed of (Kiperstok et al, 2002).

In Brazil, according to the National Solid Waste Policy (PNRS), the final disposal of industrial waste is the generator's obligation. However, there are difficulties in achieving the objectives set by the government (ABETRE, 2006). Treatment and disposal costs are obstacles for small and medium-sized industries, due to the difficulty in adopting appropriate waste treatment practices. Another obstacle is the remoteness of the treatment plant, which in some cases is located far from industrial centres and hubs (PricewaterhouseCoopers, 2006).

5.1 - Waste Classification

NBR 10004/2004 (Solid Waste) classifies solid waste according to its potential risks to the environment and public health, so that it can be managed properly. Based on its classification, the generator of a waste can easily identify its risk potential, as well as identify the best alternatives for final disposal and/or recycling.

Standard NBR 10004/2004 classifies waste as:

- Class I, hazardous - those that are dangerous and pose a risk to public health or the environment, with characteristics of flammability, corrosivity, reactivity, toxicity, pathogenicity, under the conditions established in the standard or listed in tables containing the list of hazardous waste from non-specific sources and the list of hazardous waste from specific sources.
- Class II non-hazardous:

- Class II A (non-inert) - those that do not fall into the class I or class II B classifications. They may have properties such as: biodegradability, combustibility or solubility in water.
- **Class II B (inert) -** those which, when sampled in accordance with standard NBR 10007 and solubilised in accordance with standard NBR 10006, do not have any of their constituents solubilised at concentrations higher than the water sustainability standards, with the exception of the standards for appearance, colour, turbidity, hardness and taste.

Iron and steel foundry slag is classified according to NBR 10.004/2004 as waste ranging from class II A (non-hazardous and non-inert waste) to class III (non-hazardous and inert waste), and much of this slag is stored inappropriately in open-air yards and construction sites.

5.2 - Recycling Industrial Waste

According to Marioto & Bonin (1996), the cost of generating waste affects the economy of companies in Brazil and the situation tends to worsen due to factors such as the rising costs of disposing of waste, the progressive shortage of suitable areas to deposit it and the imminent requirement to comply with international environmental standards.

The PRNS, Law 12305/10 (Brasil, 2010a) obliges large entrepreneurs to choose between waste reduction, reuse and recycling.

The reuse of materials through recycling extends the life cycle of materials, thereby reducing the consumption of non-renewable natural resources, conserving the environment and avoiding the exhaustion of certain natural reserves (Medina, 2006).

In recent years, research into the reuse of industrial waste has intensified around the world. In developed countries, recycling is seen as a highly profitable market. High investment in research generates an increase in the quality of the recycled product and leads to more effective production (Menezes et a/., 2002).

In Brazil, waste recycling still has low rates compared to the amount produced, although these figures have increased in recent decades due to the dedication of researchers to studying this issue, obtaining very relevant results (Menezes et al., 2002). Compared to developed countries, national waste reuse is still timid (Angulo et al., 2001).

The ceramics industry is one of the most prominent in the recycling of industrial and urban waste, due to its relatively simple processing techniques, the low performance required of the products, and the physical and chemical characteristics of the ceramic raw materials (Oliveira et *al.,* 2004). The use of solid waste as ceramic additives can also help to improve the properties of ceramics and facilitate their processing.

5.3 - Recycling Solid Waste and Incorporating it into Structural Ceramics.

Ferreira (2012) evaluated the use of some industrial waste in the manufacture of sealing blocks. The waste used was stone dust from granite processing, biomass boiler ash and sludge from the pulp and paper manufacturing effluent station. Systems with the addition of 10%, 20% and 30% waste by mass were studied and physical tests of linear shrinkage, absorption, loss on ignition, tensile strength and colour after firing were carried out. The results obtained indicate that only stone dust has potential for use in the manufacture of sealing blocks. The other two residues did not meet the absorption requirement and only the sewage sludge did not meet the tensile strength required by the same standard. The data gathered from this study showed that industrial waste can be used in the manufacture of sealing blocks, however, there were limitations in terms of complying with the standard corresponding to the application of ceramic sealing blocks.

Castro (2011) studied the use of manganese reduction sludge, a waste product from the electric furnaces used to produce manganese ferroalloys, as a raw material for building bricks. Levels of 0%, 2%, 5% and 10% of waste were formulated into the clay used commercially, and sintered at temperatures of 850 °C, 950 °C and 1050 °C. This study concluded that the addition of manganese sludge to the ceramic mass for the production of red ceramics proved to be highly viable from a technical point of view.

Rêgo (2010) carried out a study on the effect of adding steel slag to ceramic masses for the red ceramics industry, with the aim of improving the quality of roof tiles made from recycling this waste. The specimens were formed by extrusion and fired at temperatures of 800 °C, 850 °C, 900 °C and 950 °C with 0%, 5%, 10%, 15%, 20% and 25% steel slag content. The technological tests of linear shrinkage, water absorption, apparent porosity, apparent specific mass and flexural strength were carried out in order to detect the characteristics of the formulations studied. The results showed that it is possible to use up to 15 per cent steel slag in the ceramic mass for roof tiles.

In his work, Junior (2009) proposed the development of red ceramics from industrial waste as a recycling alternative for the companies that generate it. His work used as raw materials: sludge from water treatment plants, foundry sand, glass microspheres from blasting, acid neutralisation salts from batteries and clay. The ceramic made with 80 per cent industrial waste, of which 50 per cent was WTP sludge, had a maximum flexural strength of 10.8 MPa at 1000 °C. As the temperature rose to 1050 °C, the strength dropped to 9.6 MPa. The new ceramic showed a high mechanical strength value, which makes it possible and viable to use this waste for the production of red ceramics.

CHAPTER 6

MATERIALS AND METHODS

The materials and methodology used, as well as the equipment and characterisation techniques are presented in this chapter. The flowchart below (Figure 15) contains a general outline of the procedure used in the methodology. It is worth noting that the preparation and formulation of the clays, the granulometry, the plasticity of the basic clay and the formulations, the forming of the specimens (extrusion), the firing and the technological tests (RL, AA, PA, PF, TRF and MEA) were carried out at the Clay Technological Testing Laboratory - LETA at SENAI-DR-PI in Teresina. The tests (XRD and TG) were carried out at the Interdisciplinary Laboratory of Advanced Materials - LIMAV - UFPI. The XRF chemical analyses were carried out at the Materials Laboratory of the Federal University of Ceará.

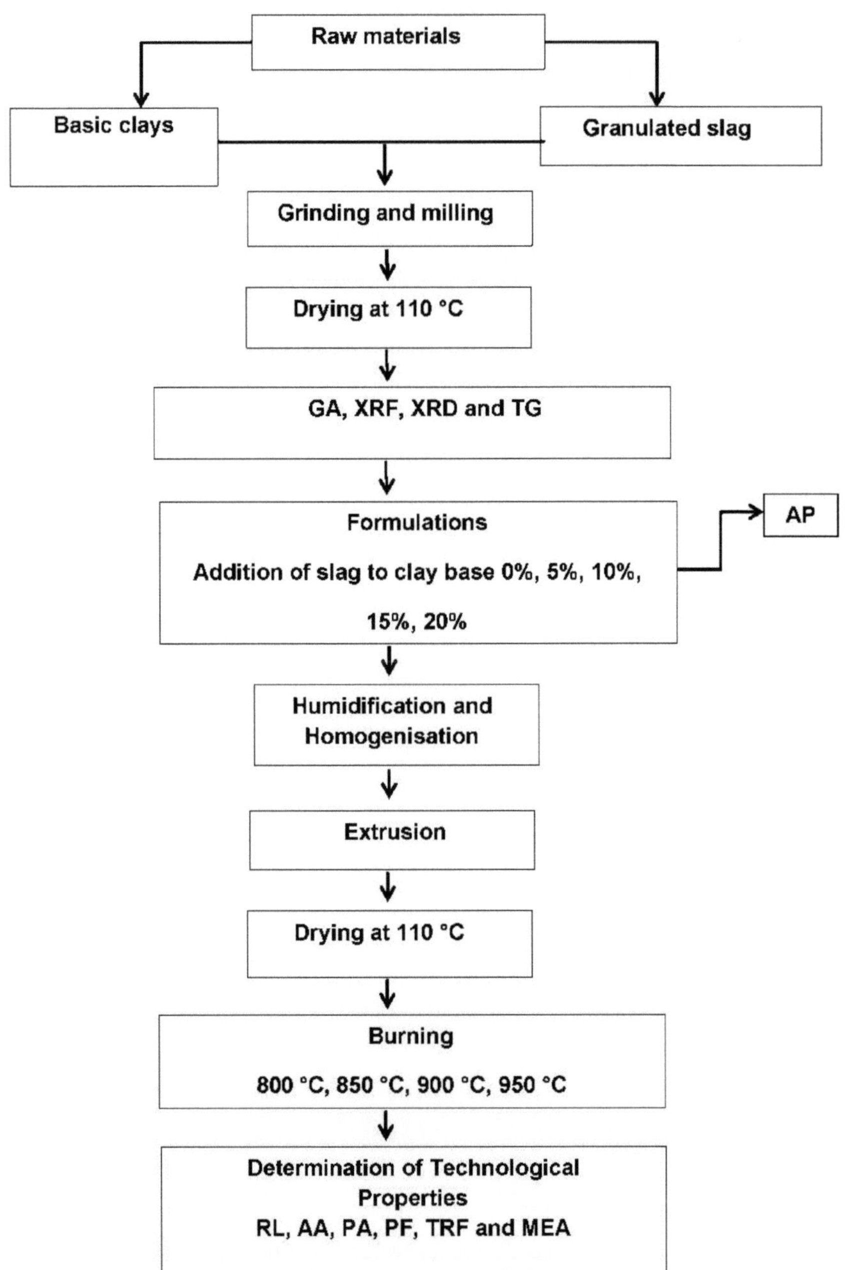

Figure 15- Flowchart of the Experimental Procedure

6.1 - Raw materials

Two raw materials were used to carry out the work: the basic mass of clays **(Figure 16 "a")** used in the ceramics industry and **granulated foundry** slag **(Figure 16 "b").** The basic mass of clays was collected from a ceramics industry in the rural area of Teresina-PI, and the slag from a foundry industry also located in the city of Teresina-PI.

Figure 16- Raw materials used in the research "a" Basic Clay Mass "b" - Granulated Foundry Slag

6.1.1 - Basic mass of clays

The basic mass of clay was collected in the courtyard of a red ceramics industry whose main products are roof tiles and bricks, in sufficient quantity to prepare the specimens and carry out the tests.

This raw material was collected in the form of clods and dried in the open air. A hammer mill with a 4 mm diameter grate was used to break it down, endeavouring to use a grain size close to that used in the ceramics industry in Piauí. A certain amount of the basic mass of clay was then separated and dried in an oven at 110 °C to carry out the chemical, mineralogical, thermal and granulometric analyses.

6.1.2 - Granulated Foundry Slag

The waste was collected based on NBR 10007:2004 for the waste under study. According to NBR 10.007:2004, the solid waste sample under study must be obtained through a sampling process that guarantees the same characteristics and properties as the total mass of the waste.

The slag was collected in the yard of the foundry industry, whose main activities are casting and machining cast iron parts. As slag is waste that is not reused in the foundry process, it is disposed of in the foundry yard without any prior treatment. The disaggregation process used was the same as that used for the basic clay mass. A hammer mill with a 4 mm diameter grid was used. The waste was then ground in a mortar until it passed through a 40 mesh sieve.

6.2 - Characterisation of Raw Materials

The raw materials were characterised using the following techniques: granulometric analysis (GA); chemical analysis by X-ray fluorescence (XRF); mineralogical analysis by X-ray diffraction (XRD); plasticity analysis (PA) and thermal analysis by gravimetry (TG), with the aim of determining their chemical, mineralogical and technological characteristics.

6.2.1 - Particle size analysis

For the particle size analysis, 100g of each raw material was taken after grinding and drying in an oven at 110 °C. The granulometric analysis of the raw materials was determined by sieving, using five ABNT sieves. The sieves were placed from top to bottom in order from the **largest opening (710** pm) to the smallest (45 pm) followed by the bottom. The sieves were then shaken with an electric shaker for 20 minutes. The material retained on each sieve and bottom was then weighed.

6.2.2 - Chemical Analysis

The chemical composition of the raw materials was determined by energy dispersive X-ray fluorescence (EDX), which qualitatively and quantitatively determines the elements present in a given sample. For this type of analysis, the Shimadzu EDX-700 X-ray Fluorescence Spectrometer was used, located at the Materials Laboratory of the Federal University of Ceará (UFCE). The semi-quantitative method was used in a vacuum atmosphere. The raw materials used were ground to a particle size of less than 325 mesh (**44** pm). The results obtained are presented in the form of the most stable oxides of the chemical elements present. Due to limitations of the method, only elements between Na (12) and U (92) on the periodic table are detected.

6.2.3 - Mineralogical Analysis

X-ray diffraction (XRD) is the most comprehensive method for determining the mineralogy of ceramic raw materials due not only to the possibility of identifying the mineral species present, but also because it makes it possible to study the crystallographic characteristics of these minerals (Barba, 1997).

The mineralogical analyses of the raw materials were obtained by XRD tests carried out with material ground down to 325 mesh (44 pm). The equipment was a Shimadzu XRD-6000 with a Cu tube (À = 1.54056 A). The phases of each raw material and final product analysed were assessed by comparing the peaks generated in the diffractogram with standard charts from the JCPDF computer program, registered with the ICDD (International Center for Diffraction Data).

6.2.4 - Plasticity Analysis

The analysis was carried out for the basic clay mass and for each basic clay-slag mass composition, in an attempt to predict the minimum amount of water that should be added to the mass so that it would have sufficient plasticity to be formed by extrusion. The plasticity limit was determined using the Casagrande method, calculated according to ABNT NBR 7180:1984.

The plasticity limit (LP) indicates the minimum amount of water that the clay or ceramic mass must contain in order to be mouldable. The liquidity limit (LL) corresponds to the maximum amount of water that the clay or ceramic mass can contain to still be mouldable. The plasticity index (PI) represents the difference between the liquidity limit and the plasticity limit.

6.2.5 - Thermal Analyses

In clay materials, thermal analyses are widely used to complement characterisation tests. The main thermal analytical techniques used on clays and non-metallic materials are thermogravimetric analysis (TG) and derivative thermal analysis (DTG).

In TG, the thermogravimetric curve reveals the changes in weight that occur during the heating of a material and which can have two causes: decomposition or oxidation. DTG provides the first derivative of the TG curve as a function of time or temperature. In this technique, the mass losses in the TG curves are replaced by peaks.

A Shimadzu TGA-51H Thermogravimetric Analyser was used for thermal evaluation of the raw materials. For both types of analysis, masses of around 15 mg were used, with a particle size of less than 325 mesh, under a synthetic air flow of 50 mL/min. The heating rate was 10 °C/min between 27 °C and 1000 °C. The results were analysed and the TG derivative curve, known as DTG, was obtained using the TA-60 computer program for thermal analysis from Shimadzu.

6.3 - Raw Material Formulations

The ceramic masses were formulated with 0%, 5%, 10%, 15% and 20% by weight of slag, as shown in Table 1:

Table 1- Formulations with different slag contents

Pasta formulation	Concentrations by weight (%)	
	Basic mass of clays	Slag
CP0	100	0
CP5	95	5
CP10	90	10
CP15	85	15
CP20	80	20

The CP0 formulation is the standard composition that will be used, i.e. without the addition of residue. A Shimadzu digital scale was used to weigh the masses. Water was then added to the formulations and each formulation was mixed until the mass was homogeneous, then placed in sealed plastic containers for a period of 24 hours to better distribute the water between the ceramic mass particles, in order to avoid significant loss of moisture.

6.4 - Making the specimens

The specimens were formed by extrusion (Figure 17), measuring approximately 15.0 cm x 2.8 cm x 1.8 cm (length, width and thickness, respectively). The masses were homogenised manually, adding quantities of water based on the calculation of the plasticity limit of each composition.

Figure 17 - Extrusion moulding of the specimens

6.5 - Drying and firing the specimens

Once the specimens had been obtained, they were measured using a Mitutoyo analogue caliper with a precision of 0.02 mm and weighed on the scale already specified; they were then dried for 24 hours at room temperature (35 °C - Figure 18) and for 24 hours in an oven at 110 °C until they reached a constant mass. After drying, the specimens were again measured for length and weight to obtain linear drying shrinkage and extrusion humidity.

Figure 18 - Drying the Specimens at Room Temperature

The test specimens were fired in a BP Engenharia TH 91D 301-000 muffle furnace. Four temperatures were used: 800 °C, 850 °C, 900 °C and 950 °C. The heating rate was 2 °C/min for each firing,

The maximum temperature threshold was 45 minutes. Cooling took place naturally, with the samples in the oven switched off, until they reached room temperature.

6.6 - Specimen Characterisation - Technological Tests

In order to assess the properties of each final product, tests were carried out on Linear Shrinkage (LS), Water Absorption (WA), Apparent Porosity (AP), Loss on Fire (LF), and Flexural Tensile Strength (FST) and Apparent Specific Mass (ASM). For the Water Absorption (WA), Linear Shrinkage (LR), Apparent Specific Mass (ASM), Apparent Porosity (AP) and Flexural Tensile Strength (FRS) tests, the specimens were used in the same dimensions as after firing.

6.6.1 - Linear Shrinkage

Linear shrinkage is the variation in the linear dimension of the specimen after drying or firing, in per cent, after being subjected to specific temperature conditions. Positive values indicate shrinkage, negative values indicate expansion. The specimens were measured using the calipers already specified, and their lengths were used to calculate the linear shrinkage during drying (Equation 1), firing (Equation 2) and total shrinkage (Equation 3). The results were obtained from the arithmetic mean of the values found in ten different specimens.

Linear drying shrinkage (DLs)

(Equation 1)

$$\% \mathbf{RLs} = \frac{(\mathbf{Ci} - \mathbf{Cf})}{\mathbf{Ci}} \times \mathbf{100}$$

In which:

RLs - Linear Shrinkage Percentage after drying.

Ci - Initial length of the specimen, measured immediately after forming.

Cf - Final length of the specimen, measured after drying at 110° C.

Linear Retraction of Firing (RLq)

(Equation 2)

$$\% \mathbf{RLs} = \frac{(\mathbf{Ci} - \mathbf{Cf})}{\mathbf{Ci}} \times \mathbf{100}$$

In which:

RLq - Linear Shrinkage Percentage after firing at temperature.

Ci - Initial length of the specimen, measured at 110 °C.

Cf - Final length of the specimen, measured after firing at temperature.

Total Linear Shrinkage (LR)

(Equation 3)

$$\%\boldsymbol{RL} = \boldsymbol{Rls} + \boldsymbol{Rlq}$$

In which:

RL - Total Linear Shrinkage Percentage

RLs - Linear Shrinkage Percentage after drying

RLq - Linear Shrinkage Percentage after firing at temperature

6.6.2 - Water Absorption and Apparent Porosity

The procedure for obtaining water absorption and apparent porosity was based on the ABNT NBR 6458:1984 standard.

The dry weight of the specimens was determined and they were then immersed in water for 24 hours, as shown in Figure 19.

Figure 19- Immersion of the specimens to analyse water absorption.

After being immersed, the excess water was removed using a clean, damp cloth to determine the wet weight and the immersed weight. With this data, water absorption and apparent porosity were calculated using the equations below:

Water absorption is measured using the equation:

(Equation 4)

$$\% \, AA = \frac{Pu - Ps}{Ps} \times 100$$

In which:

AA = Water Absorption Percentage

Pu= Weight of wet specimen

Ps= Weight of dry specimen

Apparent porosity was measured using the equation:

(Equation 5)

$$\% \, PA = \frac{Pu - Ps}{Pu - Pi} \times 100$$

In which:

AP = Percentage of Apparent Porosity

Pu= Weight of wet specimen

Ps= Weight of dry specimen

Pi = Weight of the immersed specimen.

The results for water absorption and apparent porosity were obtained from the arithmetic mean of the values found in ten different specimens.

6.6.3 - Loss on Fire

Loss on ignition is the decrease in mass of the dry sample during firing. In this work, the Loss on Fire was determined at each firing temperature (800 °C, 850 °C, 900 °C and 950 °C). The results were obtained by the arithmetic mean of the values found in ten different specimens.

The loss on ignition is calculated using the equation:

(Equation 6)

$$\% \, PF = \frac{Pi - Ps}{Pi} \times 100$$

In which:

PF - Fire Loss Percentage.

Pi - Weight of the specimen after complete drying at 110°C.

Pf - Weight of the specimen, measured after firing.

6.6.4 - Apparent Specific Mass

It is the result of the ratio between the mass of the dry specimen and its apparent volume. The apparent specific mass was calculated based on ABNT NBR 6458:1984 using the following equation:

(Equation 7)

$$MEA = \frac{Ps}{Pu - Pi} \; (g/cm^3)$$

In which:

MEA = Apparent Specific Mass

Ps = Weight of dry specimen after firing

Pu = Weight of wet specimen after firing

Pi = Weight of immersed specimen after firing

The apparent specific mass results were obtained from the arithmetic mean of the values found in ten different specimens.

6.6.5 - Bending Tensile Strength

Flexural tensile strength is the stress required to break a specimen. The mechanical strength of the specimens was assessed using the three-point bending tensile strength test, based on the ABNT NBR 13816:1997 standard. The TRF is determined using the three-point bending test, calculated using the equation below:

(Equation 8)

$$TRF \; (Kg/cm^3) = \frac{N \times P}{b \times h^2}$$

In which:

TRF = Flexural Tensile Strength

P = Load (in kgf) reached at the moment of rupture

N= Number of weights

b = Width (in cm) of the specimen

h = Height (in cm) of the specimen

The specimens were measured in all three dimensions and tested using a Shimadzu AG-I universal testing machine with a capacity of 250kN. The lowest capacity load cell (50kN) was used for this test, with a maximum load sensor of 0.5kN, in order to obtain more accurate results. The results for tensile strength and bending were obtained from the arithmetic mean of the values found in ten different specimens.

CHAPTER 7

RESULTS AND DISCUSSIONS

7.1- Characterisation of Raw Materials

7.1.1- Particle size analyses

Table 2 shows the results of the particle size analysis of the basic mass of clays and slag.

Table 2- Particle size distribution of the basic mass of clays and slag

ABNT sieve	Opening (jim)	Concentrations by weight (%)	
		Basic mass of clays	Slag
25	710	32,05	53,67
40	425	27,45	34,08
100	150	20,13	6,51
200	75	11,58	3,39
325	45	7,26	2,14
Fund	-	1,53	0,21

The data in Table 2 shows that the slag fraction has a predominantly coarser grain size than the basic clay mass. This can be explained by the cooling achieved after the casting process. A very coarse grain size can jeopardise the technological properties of structural ceramics. Finer grain size contributes to more efficient reactivity during sintering due to the increased contact surface (Roller, 1981).

7.1.2- Chemical analyses

Table 3 shows, as a percentage of mass, the chemical composition of the basic clay mass and granulated foundry slag used in this work.

Table 3- Chemical components of the basic mass of clays and slag

Raw materials	Chemical components (%)									
	SiO_2	Al_2O_3	Fe_2O_3	K_2O	MgO	CaO	TiO_2	MnO	SO_3	Others
Clay	60,09	24,84	6,17	6,02	0,98	0,59	0,82			0,49

Slag	50,80	20,46	4,70	___	10,04	10,38	0,63	2,02	0,51	0,46

The chemical composition of the basic clay paste showed the basic constituents to be SiO2 and Al_2O_3 oxides, with a significant predominance of SiO2, which is associated with quartz, kaolinite and muscovite. The high iron content (Fe2O3 - 6.17%) can give a red colour after firing, as well as reducing the refractoriness of the mass. The high potassium content (K_2O-6.02%), which is a melting component, helps to lower the sintering temperature and is essential for making low porosity ceramics. In the granulated foundry slag, SiO2 is predominant in the chemical composition, which may come from the refractory, characterising the slag as acidic. There were also predominant fractions of Al2O3, which also comes from refractory wear, and fractions of calcium oxide CaO.

7.1.3- Mineralogical Analyses

Figures 20 and 21 show the X-ray diffractograms of the basic clay mass and granulated foundry slag, respectively.

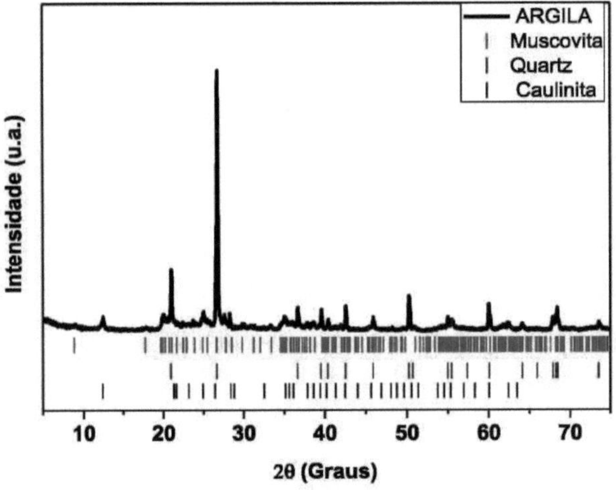

Figure 20 - X-ray diffractogram of the basic clay mass.

The diffractogram of the basic clay mass in Figure 20 shows diffraction peaks of crystalline phases referring to Muscovite ($KAl_2Si_3AlO_{10}(OH)_2$, quartz (SiO_2) and kaolinite ($Al_2Si_2O_5(OH)_4$), corroborating the XRF for the presence of silicon, aluminium and potassium oxides.

Kaolinite is the clay mineral present in kaolin and in many clays used to make ceramic products for the construction industry. This mineral is responsible for the development of plasticity in the clays studied. Quartz is the main impurity present in clays, acting as a non-plastic and inert raw material during firing. Muscovite is a mineral with a lamellar texture that can cause defects in ceramic pieces. At reduced particle size, muscovite can act as a flux due to the presence of alkaline oxides

(Vieira et al 2011).

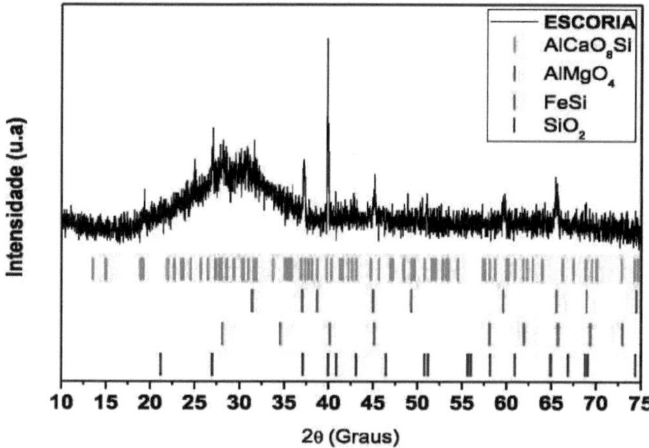

Figure 21 - X-ray diffractogram of the slag.

The diffractogram of the granulated foundry slag shows fractions of Iron/Silicon (FeSi), Quartz (SiO_2), $AlCaO_8Si$ and $AlMgO_4$. The result of the mineralogical analysis of the slag corroborates the results found in the X-ray fluorescence, in which peaks referring to quartz and iron oxide, aluminium oxide, calcium oxide and magnesium oxide were detected

7.1.4 - Plasticity Analysis

The plasticity analysis was only carried out for the basic mass of clays and for the formulations, since granulated foundry slag does not have plastic characteristics and measuring these values is not justified. Table 4 shows the plasticity indices of the basic mass of clays and the compositions studied.

Table 4 - Plasticity analysis of the basic clay mass and the formulations studied

	Plasticity Index		
Sample	**Liquidity Limit (% LL)**	**Plasticity Limit (% LP)**	**Plasticity index (% IP)**
Basic mass of clays	40,59	24,70	15,89
Basic mass of clays + 5% slag	42,31	27,29	15,02
Basic mass of clays + 10% slag	37,05	22,67	14,38
Basic mass of clays + 15% slag	34,99	21,53	13,46
Basic mass of clays + 20% slag	33,03	20,53	12,50

According to the values obtained for the basic clay mass, it can be classified as a clay mass with good plasticity (15%< IP <16%), according to SENAI (2010). The basic mass with the addition of 5% slag was also within the parameter classified as good plasticity.

It can be seen that as granulated foundry slag is added to the basic clay mix, the plasticity index decreases. This data may be related to the chemical and mineralogical composition, as well as the granulometry of the slag.

21.1.5 - Thermal Analyses

Figures 22 and 23 show the results of the gravimetric thermal analysis (TG) and differential thermal analysis (DTG) carried out on the basic clay mass and granulated foundry slag respectively.

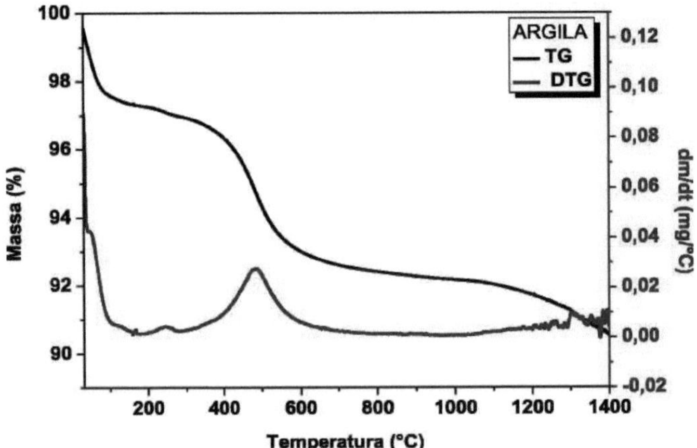

Figure 22 - TG and DTG of the basic clay mass.

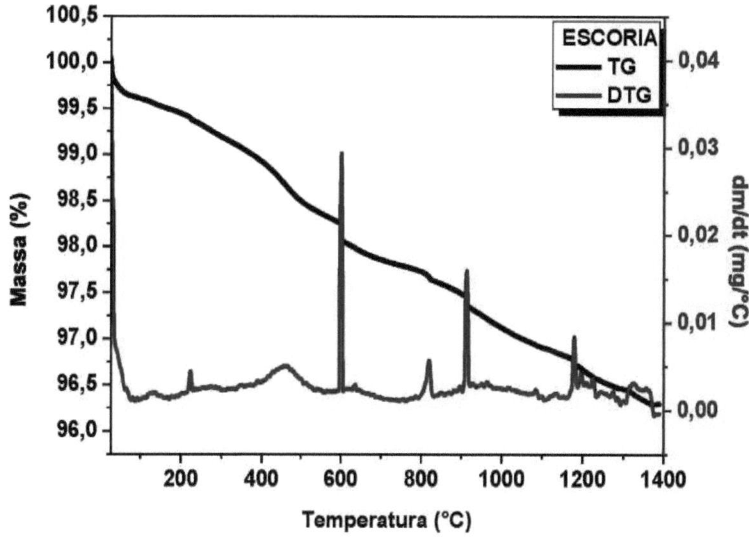

Figure 23 - TG and DTG of the slag.

The graphs show that the loss of mass in the basic clay mass is more intense than in the granulated foundry slag. This is related to the chemical composition of these materials, with the basic clay mass having a higher percentage of organic matter and water, which are easily lost at higher temperatures.

7.2 - Technological Tests

7.2.1 - Linear Drying Shrinkage (110 °C)

Figure 24 and Table 5 show the results found for linear shrinkage after drying (110 °C) of the formulations studied.

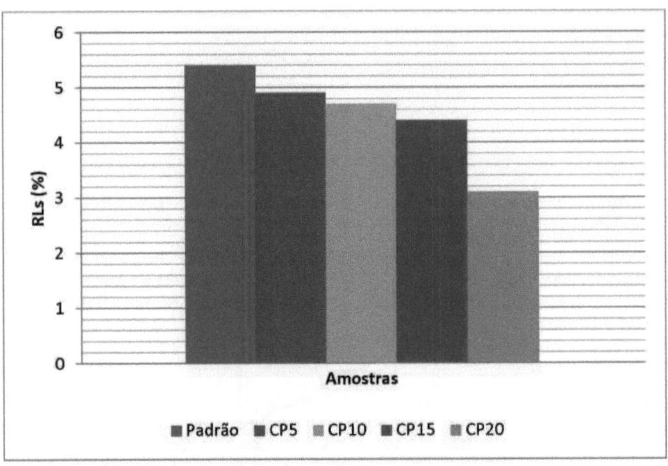

Figure 24 - Linear Shrinkage Graph for Drying at 110 °C (RLs)

Table 5 - Linear Drying Shrinkage at 110 °C

Formulations / Temperature	RLs (%)
Standard	5,4
CP5	4,9
CP10	4,7
CP15	4,4
CP20	3,1

The linear drying shrinkage data shows a downward trend with the increase in granulated foundry slag. This shows that the slag, being a deplasticiser, reduces the shrinkage of the material.

According to Pracidelli (1997), the addition of non-plastics to clays reduces their interaction with water, causing points of discontinuity in the cohesive forces between the particles. The points of discontinuity produce pores, which allow water to pass from the inside to the surface of the piece, making it easy for the clay to dry out. After tabulating the linear shrinkage data after drying the ceramic samples, we found that only the standard formulation was within the optimum range, and the other formulations were within the acceptable range, according to Dondi (2006), the optimum range (5% - 8%) acceptable range (3% - 10%).

7.2.2 - Linear Retraction of Firing (RLq)

Figure 25 and Table 6 show the results found for linear shrinkage after firing of the formulations at the temperatures studied.

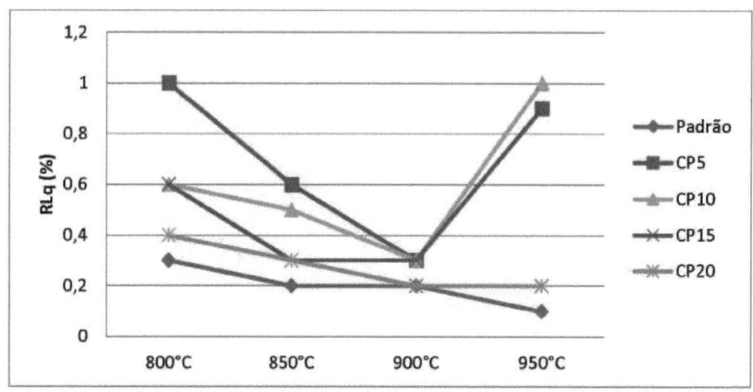

Figure 25 - Graph of Linear Shrinkage after Firing (RLq)

Table 6 - Firing retraction of the formulations at the temperatures studied

Formulations/Temperature	800 °C	850 °C	900 °C	950 °C
Standard	0,3	0,2	0,2	0,1
CP5	1	0,6	0,3	0,9
CP10	0,6	0,5	0,3	1,0
CP15	0,6	0,3	0,3	0,9
CP20	0,4	0,3	0,2	0,2

As well as shrinking as the temperature increased, the standard composition obtained the lowest linear firing shrinkage values. This behaviour is attributed to the closing of the porosity, which allows for densification of the pieces accompanied by shrinkage.

From 800 °C to 900 °C there was a decrease in the RLq in almost all the formulations, which can be explained by the occurrence of seals on the surface of the piece, which made it difficult for gases to escape from the inside of the ceramic body and, consequently, there was an expansion that reduced the shrinkage effect.

At 950 °C, due to the more pronounced sintering effect, there was an increase in the RLq. This result can be explained by the formation of a less viscous liquid phase than at 900 °C, which makes it easier to occupy the voids inside the ceramic and thus release more trapped gases.

According to Dondi (2006), the reference values for linear firing shrinkage are: optimum when less than 1.5% and acceptable between 1.5% and 3%. According to the results, the firing shrinkage was within the optimum standard for all the formulations and temperatures studied.

7.2.3 - Water Absorption

Figure 26 and Table 7 show the results found for water absorption in the formulations at the temperatures studied.

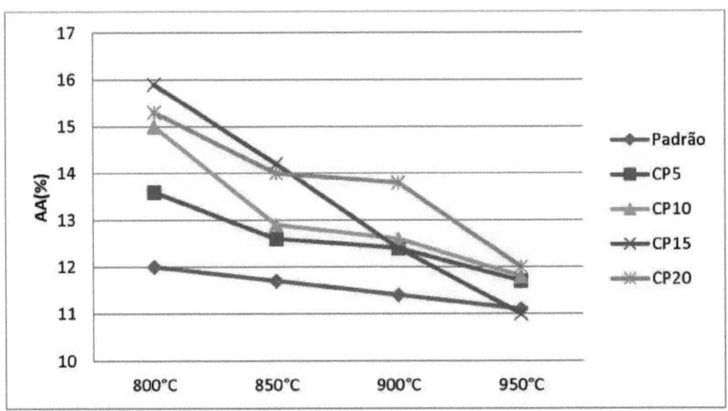

Figure 26 - Water Absorption Graph

Table 7 - Water absorption of the formulations at the temperatures studied

Formulations/Temperature	800 °C	850 °C	900 °C	950 °C
Standard	12	11,7	11,4	11,1
CP5	13,6	12,6	12,4	11,7
CP10	15	12,9	12,6	11,8
CP15	15,9	14,2	12,4	11
CP20	15,3	14	13,8	12

The compositions with granulated foundry slag had similar behaviour, with the AA data indicating a reduction with increasing temperature, with the most significant reduction at 950 °C, the temperature at which similar AA occurred in all the compositions.

All the compositions with granulated foundry slag had a higher AA than the standard composition. This behaviour is related to the greater amount of CO_2 coming out of the specimens. It can also be seen that with an increase in slag, at most temperatures, there was a small increase in water absorption and the CP5 formulation obtained the best results among the other compositions with added slag.

It is also worth emphasising that all the AA results found are in line with current ABNT standards for roof tiles and bricks. NBR 15270 and 15310 of 2005 state **that the water absorption index** must be less than 20 per cent for roof tiles, and a minimum of 8 per cent and a maximum of 20 per cent for bricks.

7.2.4 - Apparent Porosity

Figure 27 and Table 8 show the results found for apparent porosity in the formulations at the temperatures studied.

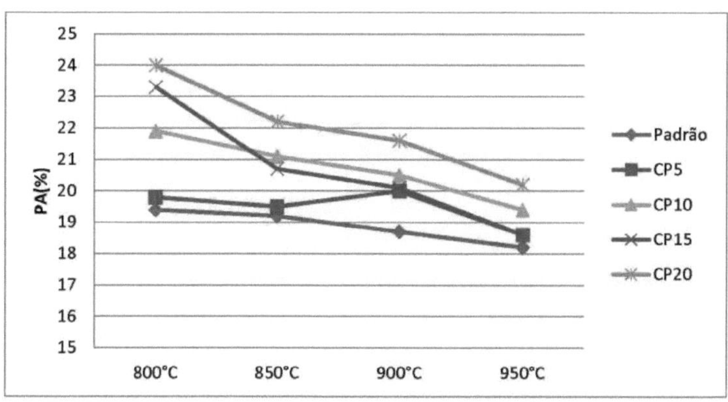

Figure 27 - Apparent Porosity Graph

Table 8 - Apparent porosity of the formulations at the temperatures studied

Formulations/Temperature	800 °C	850 °C	900 °C	950 °C
Standard	19,4	19,2	18,7	18,2
CP5	19,8	19,5	20	18,6
CP10	21,9	21,1	20,5	19,4
CP15	23,3	20,7	20,1	18,6
CP20	24	22,2	21,6	20,2

The formulations with granulated foundry slag behaved similarly to the AA results, which show a reduction with increasing temperature, with the most significant reduction at 950 °C, the temperature at which similar APs occurred in all compositions.

As can be seen from the PA results for the standard sample, there was a very rapid decrease at all the temperatures studied. These results corroborate those of the AA, in which the rapid elimination of the part's voids was verified as the temperature increased. During firing, the melting material of the standard composition promoted the formation of a low-viscosity liquid phase at a lower temperature. This liquid tends to fill the voids between the inert material in the specimen, reducing or even completely eliminating the porosity of the ceramic body.

The AP results of the compositions with granulated foundry slag are also in agreement with the AA results. The AP decreases with increasing temperature and increases with increasing granulated foundry slag in the mix; this may have been due to the increase in pores as a result of the slag itself.

The CP5 formulation obtained the best PA results among the other slag-added compositions.

7.2.5 - Fire Loss

Figure 28 and Table 9 show the results found for loss on ignition of the formulations at the temperatures studied.

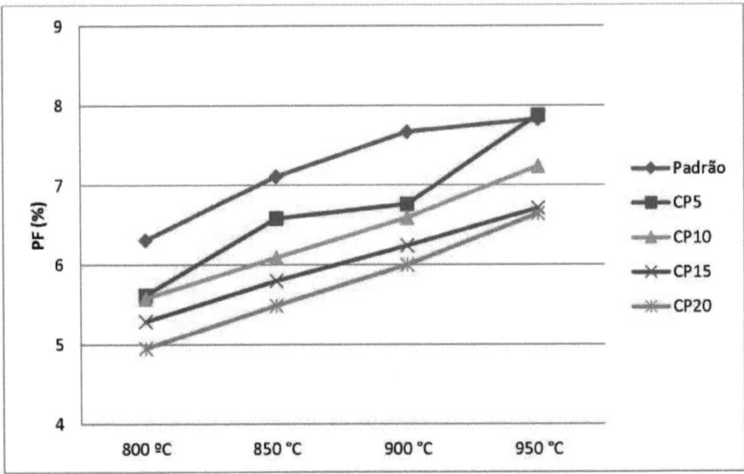

Figure 28 - Loss on Fire graph

Table 9 - Loss on ignition of the formulations at the temperatures studied

Formulations/Temperature	800 °C	850 °C	900 °C	950 °C
Standard	6,31	7,11	7,67	7,83
CP5	5,62	6,58	6,76	7,88
CP10	5,58	6,09	6,59	7,24
CP15	5,29	5,8	6,24	6,71
CP20	4,95	5,49	6	6,64

Loss on firing is the decrease in weight to a constant value, which indicates a loss of material due to the increase in temperature. It basically indicates the content of organic matter in the clay and the amount of gas and vapour formed during heating as a result of the decomposition of carbonates.

The data presented is in line with the definition of the analysis, showing that there was a gradual decrease in weight in all the formulations as the temperature increased.

It can also be seen that the standard composition exceeds the values of the compositions with the addition of granulated foundry slag. The CP20 composition showed the lowest loss on ignition percentages at all three temperatures. This can be explained by the fact that the granulated foundry slag has already been through a firing stage, whereas in the standard mass ceramic bodies, the values are more significant due to the decomposition of clay materials, as well as the combustion of organic matter.

Knowing that the ideal range of loss on ignition according to SENAI (2006) is 6% to 16%, it was found that most of the analyses were in line with the established standard.

7.2.6 - Apparent Specific Mass

Figure 29 and Table 10 show the results found for the apparent specific mass of the formulations at the temperatures studied.

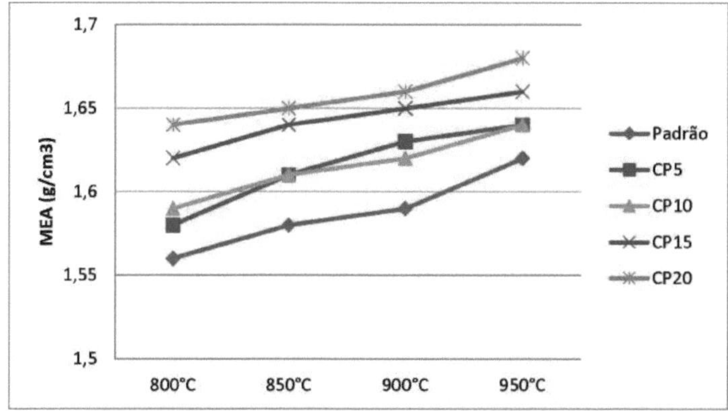

Figure 29 - Apparent Specific Mass graph

Table 10 - Apparent Specific Mass of the formulations at the temperatures studied

Formulations/Temperature	800 °C	850 °C	900 °C	950 °C
Standard	1,56	1,58	1,59	1,62
CP5	1,58	1,61	1,63	1,64
CP10	1,59	1,61	1,62	1,64
CP15	1,62	1,64	1,65	1,66
CP20	1,64	1,65	1,66	1,68

The results obtained for MEA showed an inverse behaviour to those of PA and AA. All the formulations studied showed an increase in MEA as the temperature rose. In other words, the higher the density of the ceramic body, the fewer empty spaces inside the piece. In this case, a probable predominance of sintering by viscous flow (vitrification) contributes to the densification of the pieces. The best results were obtained at 950 °C. At this temperature, the liquid phase is less viscous, which contributes to its spreading and filling of the empty spaces inside the ceramic piece.

7.2.7 - Bending Tensile Strength after Drying at 110 °C

Figure 30 and Table 11 show the results found for the drying flexural tensile strength of the formulations at 110 °C.

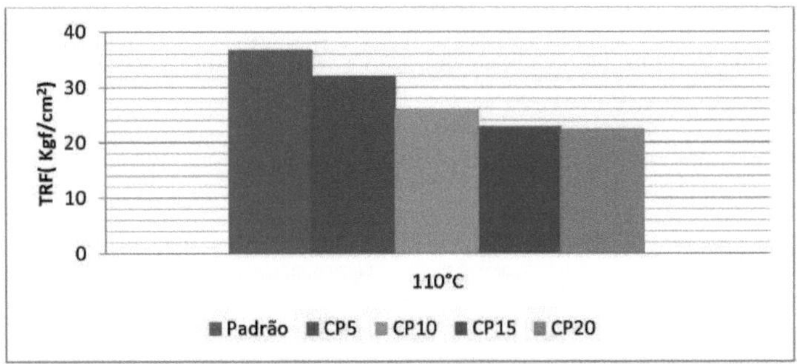

Figure 30 - Flexural tensile strength graph after drying at 110 °C

Table 11 - Flexural Tensile Strength after Drying at 110 °C

Formulations/Temperature	110 °C
Standard	36,75
CP5	32,1
CP10	26,02
CP15	22,86
CP20	22,44

According to the TRF results after drying (110 °C), there was a reduction in its value as the granulated foundry slag content increased. This may have been due to the nature of the slag itself, as it is a de-plasticiser, due to less packing in the moulding of the specimens caused by the greater quantity of coarse granules from the slag.

All the formulations had a TRF after drying within the minimum established for bricks, which according to Santos (1989) is 15 kgf/cm^2. For roof tiles, only the basic mass and the CP5 formulation reached the minimum, which is 30 kgf/cm^2.

7.2.8 - Flexural tensile strength after firing

Figure 31 and Table 12 show the results found for the firing flexural tensile strength of the formulations at the temperatures studied.

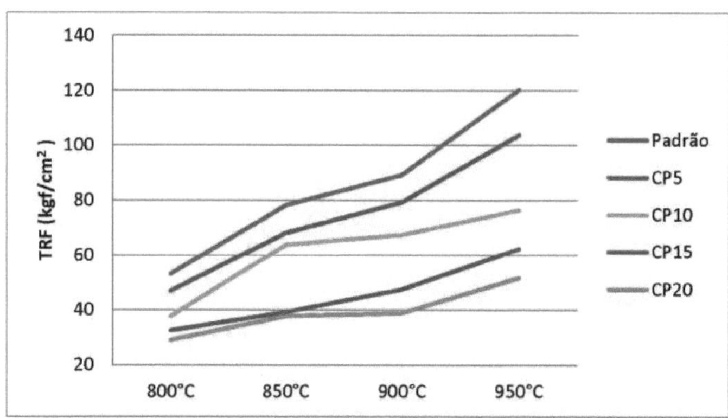

Figure 31 - Burning Flexural Tensile Strength Graph

Table 12 - Flexural tensile strength after firing the formulations at the temperatures studied

Flexural Tensile Strength after Firing - TRF (Kgf/cm^2)				
Formulations/Temperature	**800 °C**	**850 °C**	**900 °C**	**950 °C**
Standard	53,12	78,24	89,19	120,3
CP5	46,94	68,1	79,2	103,65
CP10	37,73	63,84	67,4	76,4
CP15	32,55	39,22	47,44	62,27
CP20	29,02	37,79	38,86	51,73

The mechanical strength of the standard composition and the four formulations of granulated foundry slag increased as the temperature rose. The increase in strength with increasing temperature is due to the reduction in porosity and the microstructure formed in the sintering process, such as the glassy phase, which provides greater strength to the ceramic body.

The TRF results of the fired specimens showed a reduction as the granulated foundry slag content increased at all the temperatures studied. These results are due to the coarser grain size of the granulated foundry slag, which significantly reduces the mechanical strength of the ceramic.

NBR 15270 and 15310 of 2005 require that the minimum firing TRF be 100 Kgf/cm^2 for roof tiles and 15 Kgf/cm^2 for bricks. Analysing the data found in the research for TRF after firing, it was found that all the formulations are within the standards for bricks, and only the Standard and CP5 formulations at 950 °C are within the reference standards for roof tiles.

7.3 - Macrostructural Analysis of the Burned Specimens

Figures 32, 33, 34, 35 and 36 show the macro-structural analysis of the test specimens of all the compositions (CP0, CP5, CP10, CP15, CP20) as a function of firing temperature.

It can be seen that in all compositions and firing temperatures there were no defects due to the presence of slag, and none of them affected the aesthetics of the ceramic body.

Although granulated foundry slag has a high iron oxide content, no change was observed in the colour of the ceramic bodies.

Figure 32 - Test specimens in the Standard composition

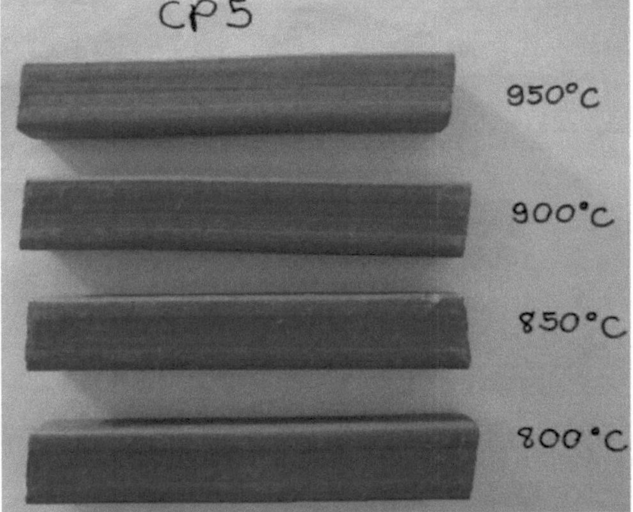

Figure 33 - CP5 specimens

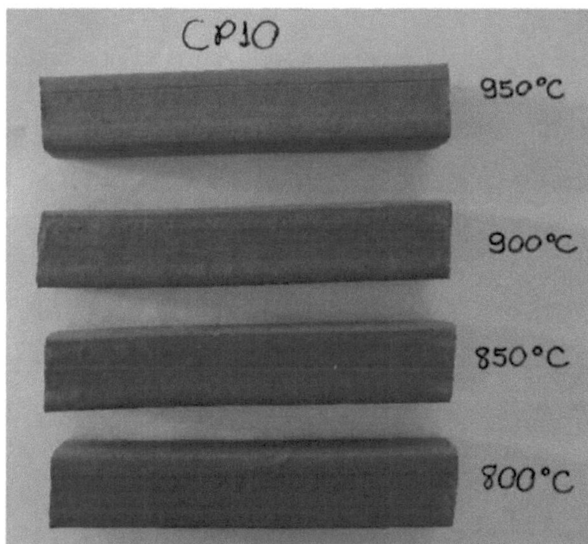

Figure 34 - CP10 specimens

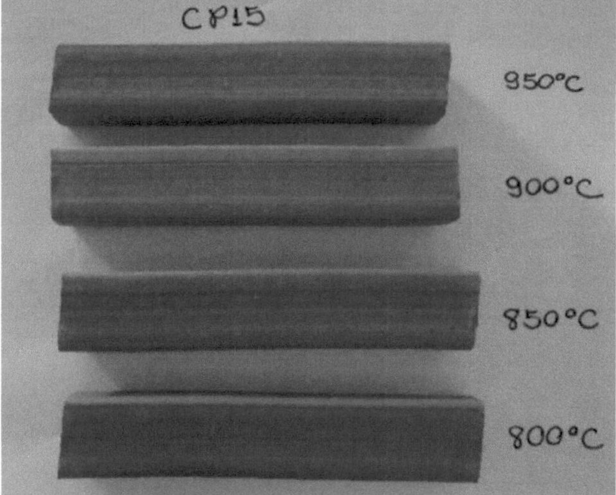

Figure 35 - CP15 specimens

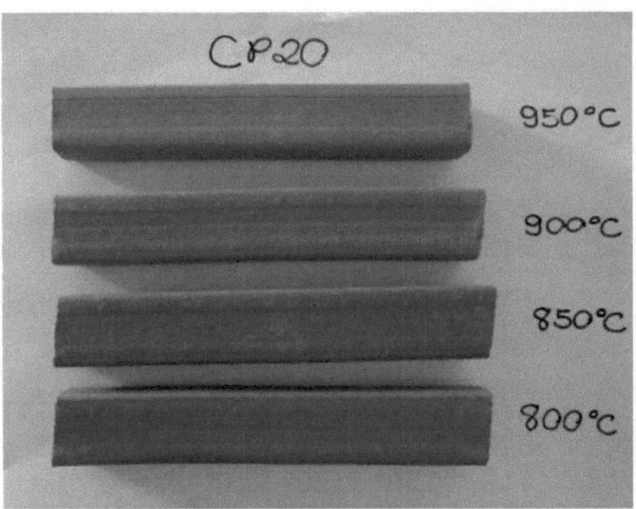

Figure 36 - CP20 specimens

CHAPTER 8

CONCLUSION

The results showed that the use of granulated foundry slag in the production of structural ceramics is a viable alternative for maintaining the standards required by structural ceramics, since most of the basic clay and slag formulations remained within the required parameters at the temperatures studied.

Another advantage of using this waste in structural ceramics is that it contributes to the planet's environmental quality, reducing the environmental impacts that can be caused by this waste.

The standard formulations and CP5 obtained the best results in the technological tests carried out, with CP5 being the best basic mass composition of clays and slag to be used in structural ceramics.

The linear shrinkage data after drying the ceramic samples showed that all the formulations were within the acceptable range, and that the firing shrinkage was within the optimum standard for all the formulations and temperatures studied.

It is also worth emphasising that all the results found for AA are in line with the current standards regulated by ABNT for roof tiles and bricks. PA had similar results to AA, decreasing as the temperature increased and increasing as the crystallised foundry slag in the mixture increased; this may have been due to the increase in pores as a result of the slag itself.

All the formulations had the TRF after drying within the minimum established for bricks, for roof tiles, only the CP5 formulation reached the minimum, which is 30 kgf/cm^2. The firing TRF analyses showed that all the formulations are within the standards for bricks and only the standard and CP5 formulations are within the reference standards for roof tiles at a temperature of 950 °C.

BIBLIOGRAPHICAL REFERENCES

ABC - Brazilian Ceramics Association. **Technical information - definition and classification.** Brazil, 2002. Available at: http://www.abceram.org.br. Accessed on: May 20013.

ABC - Brazilian Ceramics Association. Ceramics in Brazil - **sector figures - red ceramics.** sector. Brazil, 2002. Available at: http://www.abceram.org.br. Accessed: May 2013.

ABC - Brazilian Ceramics Association. Ceramics in Brazil - **Technical information: manufacturing process - flow chart 1 - red ceramic manufacturing process** - 20001b. Available at:

<http://www.abceram.org.br/asp/fg01.asp>. Accessed March 2013

ABC - Brazilian Ceramics Association. **Technical information: manufacturing processes**. 2011b. Available at: <http://www.abceram.org.br/ site/index.php?area=4&submenu= 50>. Accessed in March 2013.

ABETRE - Brazilian Association of Waste Treatment Companies. **Profile of the waste treatment and environmental services sector**. São Paulo: Abetre, 2006.

ABIFA - Brazilian Foundry Association. **Yearbook 2010, edition 121, 2011.** Available at < http://www.abifa.org.br/>. Accessed June 2013

ABIFA - Brazilian Foundry Association. Activities of the installed study commissions. **Revista Fundição & Matérias-primas**, 113ª ed., São Paulo, October, 2009.

ABIFA - Brazilian Foundry Association. Foundry waste: a solution on the way. **Magazine Fundição & Matérias-primas**. 95ª ed. São Paulo, March 2008.

ABNT - Brazilian Association of Technical Standards, **NBR 8497: Test for Determining Water Absorption of Blocks and Bricks.** Rio de Janeiro, 1984.

ABNT - Brazilian Association of Technical Standards: **NBR 6458 - Soil - Determination of the Liquidity Limit**, Rio de Janeiro, 1984.

ABNT - Brazilian Association of Technical Standards: **NBR 7180 - Soil - Determination of the Plasticity Limit**, Rio de Janeiro, 1984.

ABNT - Brazilian Association of Technical Standards: **NBR 13816. Ceramic tiles for cladding**, Rio de Janeiro, 1997.

ABNT- Brazilian Association of Technical Standards- **NBR 10004: Solid waste classifications**, Rio de Janeiro, 2004.

ABNT- Brazilian Association of Technical Standards- **NBR 10007: Sampling of Solid Waste**, Rio de Janeiro, 2004

ABNT- Brazilian Association of Technical Standards- **NBR 15270: Ceramic blocks for masonry, specification**, Rio de Janeiro 2005.

ABNT- Brazilian Association of Technical Standards- **NBR 15310: Roof tiles, specification**, Rio de Janeiro 2005

AMERICAN FOUNDRY SOCIETY. **43rt Census of World Casting Production** - 2008. Modern Casting, Illinois, p.17-21, dec 2009.

Ângulo, S. C.; Zordan, S. E.; John, V. M. **Sustainable development and the recycling of waste in the construction industry**. In: IV Seminário Desenvolvimento sustentável e a reciclagem na construção civil - materiais reciclados e suas aplicações. São Paulo-SP, 2001.

ANICER - National Ceramic Industry Association. Available at: <http://www.anicer.com.br>. Accessed on: May 2013.

Assunção, F. C. R. et. *al.* **Foundry Sector Study 2004 - 2006.** The Foundry Sector in Brazil: productive and technological profile**.** Rio de Janeiro, 2007.

Bauer, L. A. F. **Materiais de construção.** 5. ed. Rio de Janeiro: Livros Técnicos e Científicos, 2000. 705 p.

Beltran, V.; Ferrando, E.; García, J.; Sánchez, E. Extruded Rustic Floor Tile I. **Impact of the Composition on the Body's Behaviour in the Prefiring Process** Stages, **Tile & Brick Int**. v. 11, n. 3, p. 169-176, 1995.

Bethell, L. (org.) **History of Latin America.** Vol. V: from 1870 to 1930. 161 Translated **by Geraldo Gerson de Souza. São Paulo: Edusp; Imprensa Oficial SP; Brasília**, DF: Funag, 2002

Bezerra, F. D. et al . **Profile of the red ceramics industry in the Northeast**. In : 44 °Congresso Brasileiro de Cerâmica, 2005, São Paulo-SP.

Bidone, F.R.A. **Metodologia e técnicas de minimização, reciclagem e reutilização de resíduos**

sólidos urbanos. Rio de Janeiro: ABES (Association of Sanitary and Environmental Engineering), 1999.

BNB- Banco do Nordeste do Brasil. **Red ceramics sector report**. 2010. Available at:

<http://www.bnb.gov.br/content/aplicacao/etene/etene/docs/ano4_n21_informe _sectorial_ceramica_vermelha.pdf> Accessed March 2013.

Brazil - National Environment Council (CONAMA). **Resolution no. 313 of 29 October 2002**, on the National Inventory of Solid Industrial Waste.

Brazil. **National solid waste policy**. Law No. 12.305/2010. Brasília, 2010a. Available at: <https://www.planalto.gov.br/ccivil_03 /_ato2007- 2010/2010/lei/l12305.htm>. Accessed in July 2013.

Campos Filho, M. P. **Solidification and casting of metals and their alloys**. Rio de Janeiro: Technical and Scientific Books, 1978.

Carmelio, J. S. et al. **ABIFA Foundry Guide: 2009 Yearbook**. Brazilian Foundry Association. São Paulo: ABIFA, 2009.

Carmelio, J.S. et al. **ABIFA Foundry Guide**: 2011 Yearbook. Brazilian Foundry Association. São Paulo: ABIFA, 2011.

Carnin, R. L. P. **Reusing green foundry sand waste as aggregate in asphalt mixtures**, 2008. Thesis (Doctorate in Chemistry), Federal University of Paraná, Curitiba.

Caspers, K.H. Melting synthetic cast iron in a cubilot furnace. **Revista Fundição e Serviços**, São Paulo, March, p.34 - 44,1999.

Castro, C. G. **Estudo do aproveitamento de rejeitos do beneficiamento do manganês pela indústria cerâmica**, 2011. Dissertation (Master's in Materials Engineering) - Federal University of Ouro Preto, Ouro Preto.

CEPRO - Foundation for Economic and Social Research in Piauí. **Diagnosis and guidelines for the**

mineral sector in the state of Piauí. Teresina - PI: CEPRO Foundation, 2005.

CETEM-Centre for Mineral Technology. **Clay for Red Ceramics**, 2008. Available at < http://www.cetem.gov.br/publicacao/CTs/CT2008-185-00.pdf. Accessed January 2013

Chegatti, S. **Aplicação de resíduos de fundição em massa asfaltica, cerâmica vermelha e fritas cerâmicas**, 2004. Dissertation (Master's in Environmental Engineering), Federal University of Santa Catarina, Florianópolis, 2004.

Correia, S.L. **Desenvolvimento de Metodologia de Formulação de Massas Cerâmicas Triaxiais Utilizando Delineamento de Misturas e Otimização**, 2004. Thesis (Doctorate in Materials Science and Engineering) - Federal University of Santa Catarina, Florianópolis.

da Silva, A. L. **Recycling of crystallised slag for the production of mortar**, 2006. Dissertation (Master's in Materials for Engineering)- Federal University of Itajubá, Itajubá.

Dondi, M. Technological Characterisation of Clay Materials: Experimental Methods and Data Interpretation. **Revista Cerâmica Industrial**, v.11 n°3, May/June, pág. 36-40, 2006.

Eriksson, K. Environmental aspects from a foundry perspective. In: International Conference: Foundry waste possibilities in the future, San Sebastian, 2001. **Proceedings.** San Sebastian, Spain: 2001.

Facincani, E. **Tecnologia cerâmica** - I laterizi. Italy, Gruppo Editoriale Faenza Editrice. Faenza. Second edition. 267p, 1992.

Fagundes, A. B. **Mapping the management of green foundry sands in the State of Paraná from the perspective of Cleaner Production: A Contribution to the Establishment of Strategies.** 2010. Dissertation (Master's in Production Engineering). Industrial Management, Research and Postgraduate Management, Ponta Grossa Campus, Paraná.

Ferreira, L.C. **Potential use of industrial waste in the formulation of red ceramic mass for the manufacture of sealing blocks.** 2012. Dissertation (Urban and Industrial Master's Degree) - Federal University of Rio Grande do Paraná, Rio Grande do Paraná.

Gorbea, I.S. Activities carried out in spain with reference to the environment and foundries. In: International Conference: Foundry waste possibilities in the future, San Sebastian, 2001. **Proceedings.** San Sebastian, Spain: 2001.

Grun, E. **Characterisation of clays from Canelinha/SC and study of the formulation of ceramic masses**. 2007. Dissertation (Master's in Materials Science and Engineering) - Santa Catarina State University, Joinville.

Grun, E. et al. **Definition of Parameters for the Formulation of Red Clay Mixtures**, In: 49th Brazilian Ceramics Congress, São Pedro, SP, 2005

INT - National Institute of Technology. **Panorama of the Red Ceramic Industry in Brazil**, Rio de Janeiro, 2012

Junior, V.M.T. **Development of new ceramic materials from water treatment plant sludge, glass microbeads from blasting, battery acid neutralisation salts and foundry sand**, 2009. Dissertation (Master's in Engineering). University of Paraná.

Kalyoncu, R.S. Slag - Iron and steel. U.S. **Geological Survey Minerals Yearbook-** 2000. Available at: <http://minerals.usgs.gov/minerals/pubs/commodity/iron_&_steel_slag/790400. pdf>. Accessed on: 15/03/2006

Kiperstok, A. et al. **Pollution Prevention**. Brasília: CNI/SENAI.ch.6, p. 183 to 221, 2002.

Kondic, V. **Metallurgical principles of casting**. São Paulo: Polígono, 1973.

Loper, J.R.; Carl, R. Castirons - Essential alloys for the future. In: LXV World Foundry Congress, South Korea, Oct. 2002. **Foundryman,** v. 96, part 11, nov. 2003.

Mahan, W. Slag Control. In: **Cupola handbook** 5 ed. U.S.A: American **Foundrymen's Society, Des Plaines, Illinois** 60016, 1994. Chap 20 p. 249-259.

Manfredini & Schianchi. **Plants for Red Ceramics and Extruded Ceramic Products**. Manfredini & Schianchi Equipment. 2009. Available at

<http://www.manfredinieschianchi.com/208-4PO-plantas-para-ceramica red-e-for-products-extruded-from-ceramics.htm>. Accessed on 14 October 2009.

Marioto C.; Bonin A. L. Treatment of sand waste. **Foundry and raw materials,** Mar/Apr (1996).

Medina, H. V. Sustainable Production and Use of Materials: Environmental Management and Life Cycle Analysis. In: 61st Annual Congress of the ABM - Brazilian Association of Metallurgy and Materials, 2006, Rio de Janeiro. **Proceedings** of the 61st ABM Annual Congress. São Paulo, Brazil, 2006. v. 02. p. 1781-1790.

Menezes, R. R.; Neves, G. A.; Ferreira, H. C. O estado da arte sobre o uso de resíduos como matérias-primas cerâmicas alternativas. **Revista Brasileira de Engenharia Agrícola e Ambiental**, Campina Grande, v. 6, n. 2, p. 303-313, 2002.

Ministry of Mines and Energy. **Statistical Yearbook: Non-metallic transformation sector.** Department of Geology, Mining and Mineral Transformation. Brasilia: 2009.

Montecelli, C. A. **The competitiveness of the Brazilian foundry industry**. 1994. Dissertation (Master's in Economics) - State University of Campinas. Campinas

Monteiro, C. M. O. L. **Influence of gypsum on the appearance of efflorescence in ceramic tiles** 2009. Dissertation (Master's in Materials Science and Engineering) Federal University of Rio Grande do Norte, Natal.

Moraes, C.A.M. **Ferros e steels fundidos**. Porto Alegre-UFRS, 2000.

Morais, D. M, **Ligno-cellulosic Waste Briquettes as Energy Potential for Firing Ceramic Blocks**: Application in an Industry Supplying the Federal District, 2007. Thesis (Doctorate in Structures and Civil Construction) - Department of Civil and Environmental Engineering, University of Brasília, Brasília.

Morais, D. M.; Sposto R. M., Technological and Mineralogical Properties of Clays and their Influences on the Quality of Ceramic Fence Blocks Supplying the Federal District Market. **Cerâmica** industrial, São Paulo, v. 11, n. 5, p.35-38, 2006.

Motta, J. F. M.; Zanardo, A.; Cabral, M.J. As matérias-primas cerâmicas. Parte I: O perfil das principais indústrias cerâmicas e seus produtos, **Cerâmica Industrial**, v. 6, p. 29-31, March/April, 2001.

Motta, J.F.M.; Zanardo, A.; CabraL JR., M.; Tanno, L. C. and Cuchierato, G.. Plastic raw materials for traditional ceramics: clays and kaolins. **Cerâmica Industrial**, vol.9, n°2, p. 33-46, 2004.

Naystron, P. Use of bentonite bonded foundry sand and slag from cupola furnaces in landfill construction. In: International Conference: Foundry waste possibilities in the future, San Sebastian, 2001. **Proceedings.** San Sebastian, Spain: 2001.

Norton, F. H. **Introduction to Ceramic Technology.** Editora Edgar Blucher Ltda. Editora da Universidade de São Paulo. São Paulo, 1973.

Oliveira, A. P. N. et al. Raw materials used in the manufacture of bricks and building blocks: characteristics and influence on the properties of the final product, **Cerâmica Informação**, v. 10, p.57-65, 2000.

Oliveira, G. E.; Holanda, J. N. F. 2004. Reusing solid waste from the steel industry in red ceramics. **Cerâmica**, Campos dos Goytacazes-Rio de Janeiro, v. 50, p. 75-80, 2004.

Oliveira, K. R. B. **Evaluation of ceramic blocks produced for the metropolitan region of Goiana - GO**. 2002. Dissertation (Master's in Civil Engineering) - Federal University of Goiás, Goiânia.

Oliveira, T. M. N. **Eco-strategia empresarial no setor metal-mecânico da escola técnica Tupy**, 1998. Thesis (Doctorate in Production Engineering), Federal University of Santa Catarina, Florianópolis.

Pauletti, M. C. **Model for the introduction of new technology in clusters of micro and small companies: a case study of the Red Ceramics industry in the Tijucas river valley.** 2001. Dissertation (Master's Degree in Engineering

Production) - Federal University of Santa Catarina, Florianópolis.

Pérez, C. A. S.; Paduani, C.; Ardisson, J. D.; Gobbi, D.; Thomé, A. Characterisation of ceramic masses used in the red ceramics industry in São Domingos do Sul - RS. **Cerâmica Industrial**, São Paulo, v. 15, n° 1, p. 38-43, 2010.

Petrucci, E.G. R. **Construction Materials**. 11ª .ed. Porto Alegre: Globo, 435p 1998.

PWC- Princewatherhousecooprs. Study on the industrial waste treatment sector, 2006.

Rêgo, V.R. **Efeito da adição de escória de aciaria em formulações de massa cerâmica para telha.** 2010. Thesis (Doctorate in Materials Science and Engineering) Federal University of Rio Grande do Norte, Natal.

Ribeiro, D. V., Morelli, M. R., **Solid waste - problem or opportunity?** Rio de Janeiro: Ed. Interciência, 2009.

Ribeiro, R.A.C. **Development of new ceramic materials from industrial metal-mechanical waste**. 2008. Dissertation (Master's in Engineering) - Federal University of Paraná, Curitiba.

Roller, P.W.. Granulation of iron ore sinter feeds. **BHP Technical Bulletin**, v 25, n° 1, May, p.79, 1981.

Rossitti, S. M. **Casting processes and variables**. Grupo Metal, May 1993

Schwob, . Prospects for the diffusion of natural gas in the Brazilian red ceramics industry, 2007. Dissertation (Master's degree in energy planning) - Federal University of Rio de Janeiro, Rio de Janeiro.

SEBRAE - MG. Brazilian Micro and Small Business Support Service. Red ceramics Strategic sectors - **The red ceramics segment - 2005.** Available at: <http://www.sebraemg.com.br>. Accessed in March 20013.

SEBRAE. Brazilian Micro and Small Business Support Service. **Red ceramics: market study SEBRAE/ESPM** 2008: full report. [Available at: <http://www.biblioteca.sebrae.com.br/bds/bds.nsf/C5B4284E12896289832574 C1004E55 DA/$File/NT00038DAA.pdf>. Accessed in February 2013.

SENAI - National Industrial Apprenticeship Service - Piauí Regional Department. **Special Programme in Red Ceramics Technology. Teresina - PI,** 2006.

SENAI - National Industrial Apprenticeship Service. Piauí Regional Department. **Bases for the Competitive Improvement of Small Ceramic Pottery Enterprises**. Teresina - PI, 2006.

SENAI - National Industrial Apprenticeship Service. Regional Department of Piauí. **Ceramic Tests** - Teresina - PI, 2010...

Siegel, M. Processos de Fundição: generalidades, considerações gerais sobre a escolha do processo, importância relativa dos diversos processos. **Foundry**, 10ª ed., Brazilian Metals Association - ABM, 1978.

Silva, A. V. **Analysis of the ceramic brick production process in the state of Ceará - from raw material extraction to manufacture.** 2009. Dissertation (Master's in Civil Engineering), Federal University of Ceará, Fortaleza.

SINDICER / PI- Sindicato da INDÚSTRIA CERÂMICA PARA CONSTRUÇÃO DO PIAUÍ (Data on ceramics in Piauí). Teresina, 2008 (typed).

Soares, R. A. L. **Influence of limestone content on the physical, mechanical and microstructural behaviour of structural ceramics.** 2008. Dissertation (Master's in Materials Science and Engineering) - Federal University of Rio Grande do Norte, Natal.

Vallina, J.J. **Coke oven**. Material prepared for the SENAI cubicle oven course, 1998

Van Vlack, L. H. **Properties of Ceramic Materials**, Editora Edgard Blucher, São Paulo, p.318, 1973.

Varela, Márcio Luís. **Development of a methodology for rational mineralogical analysis of clay minerals**. 2004. Dissertation (Master's Degree in Mechanical Engineering) - Postgraduate Programme in Mechanical Engineering, Federal University of Rio Grande do Norte, Natal.

Vieira, C. M. F.; Pinheiro, R. M. Evaluation of kaolinitic clays from Campos dos Goytacazes used to manufacture red ceramics. **Cerâmica [online]**, v.57, n. 343, p.319-323, 2011.

Vieira, C. M. F.;Feitosa, H. S.; Monteiro, S.N. Avaliação da Secagem de Cerâmica Vermelha através da Curva de Bigot, **Cerâmica Industrial**, January/February, 2003

Vieira,C. M. F.;. Monteiro, S. N., **Matéria 14**, v 3, P.881-905, 2009.

Wender, A.A.; Baldo, B.B. The potential use of a clay residue in the manufacture of ceramic tiles - Part II. **Cerâmica Industrial**, São Paulo, v.3, n.1-2, p.34-36, 1998.

Printed by Books on Demand GmbH, Norderstedt / Germany